E.A. Volkov

T0229665

Block Method for Solving the Laplace Equation and for Constructing Conformal Mappings

CRC Press
Taylor & Francis Group
Boca Raton London New York

CRC Press is an imprint of the
Taylor & Francis Group, an **informa** business

First published 1994 by CRC Press
Taylor & Francis Group
6000 Broken Sound Parkway NW, Suite
300 Boca Raton, FL 33487-2742

Reissued 2018 by CRC Press

© 1994 by Taylor & Francis
CRC Press is an imprint of Taylor & Francis Group, an Informa business

No claim to original U.S. Government works

This book contains information obtained from authentic and highly regarded sources. Reasonable efforts have been made to publish reliable data and information, but the author and publisher cannot assume responsibility for the validity of all materials or the consequences of their use. The authors and publishers have attempted to trace the copyright holders of all material reproduced in this publication and apologize to copyright holders if permission to publish in this form has not been obtained. If any copyright material has not been acknowledged please write and let us know so we may rectify in any future reprint.

Except as permitted under U.S. Copyright Law, no part of this book may be reprinted, reproduced, transmitted, or utilized in any form by any electronic, mechanical, or other means, now known or hereafter invented, including photocopying, microfilming, and recording, or in any information storage or retrieval system, without written permission from the publishers.

For permission to photocopy or use material electronically from this work, please access www.copyright.com (http://www.copyright.com/) or contact the Copyright Clearance Center, Inc. (CCC), 222 Rosewood Drive, Danvers, MA 01923, 978-750-8400. CCC is a not-for-profit organiza-tion that provides licenses and registration for a variety of users. For organizations that have been granted a photocopy license by the CCC, a separate system of payment has been arranged.

Trademark Notice: Product or corporate names may be trademarks or registered trademarks, and are used only for identification and explanation without intent to infringe.

A Library of Congress record exists under LC control number: 94009091

Publisher's Note
The publisher has gone to great lengths to ensure the quality of this reprint but points out that some imperfections in the original copies may be apparent.

Disclaimer
The publisher has made every effort to trace copyright holders and welcomes correspondence from those they have been unable to contact.

ISBN 13: 978-1-138-10496-9 (hbk)
ISBN 13: 978-1-138-55779-6 (pbk)
ISBN 13: 978-1-315-15032-1 (ebk)

Visit the Taylor & Francis Web site at http://www.taylorandfrancis.com and the CRC Press Web site at http://www.crcpress.com

Preface

This book is the result of my many years of investigations carried out in the field of approximate methods of solving differential equations. Chapter 1 contains the exposition and substantiation of my own block method of solving a mixed boundary-value problem, including the Dirichlet problem and the Neumann problem for Laplace's equation on an arbitrary polygon. The boundary values of the required solution or of its normal derivative are defined on the sides of the polygon by analytic functions, in particular, by algebraic polynomials. An approximate solution of the boundary-value problem is sought on a finite number of fixed intersecting subdomains (sectors, half-disks, disks), called basic blocks, in the form of harmonic functions which are elementary under the boundary conditions specified by algebraic polynomials.

The block method is constructed with the use of integral representations of harmonic functions of the type of Poisson's integral on extended blocks which are sectors, half-disks and disks (containing the basic blocks). In these representations the corresponding integrals are approximated by the quadrature formulas of rectangles. The resulting system of algebraic linear equations is well conditioned and its solution can be found by means of simple iterations which converge in a geometric progression with a common ratio independent of the discreteness parameter. The approximate solution of a boundary-value problem, found by means of the block method, is exponentially convergent (on the basic blocks) with respect to the discreteness parameter up to the boundary of the polygon even if the defined boundary values of the solution are discontinuous at the vertices of the polygon.

In Chapter 2, the problem of approximate conformal mapping of multiply-connected polygons onto canonical domains (onto a plane with cuts along parallel segments and onto a ring with cuts along the arcs of concentric circles) is constructively solved with the use of the block method by means of a reduction to the Dirichlet problem. No preliminary information is required concerning the unknown conformal invariants of the domains being mapped. The conformal invariants are determined in the process of constructing of the approximate conformal mapping with an error decreasing exponentially with respect to the discreteness parameter. The approximate conformal mapping is obtained on the (basic) blocks in the form of elementary analytic functions.

Chapter 3 is devoted to the development and application of the block method to the approximate conformal mapping of simply-connected domains onto a strip and of doubly-connected domains onto a ring. The algorithms of constructing approximate conformal mappings of a number of specific domains are described in detail. The class of domains being mapped is considerably extended and the apparatus of the block method is developed by means of introduction of special blocks, the use of analytic continuation, and change of a complex variable. Very accurate results of calculations carried out on the BESM-6 computer according to the indicated algorithms are given which substantiate by practical computations the exponential convergence of the block method and its strong stability as concerning the rounding-off errors. All the computations were carried out by me.

In Chapter 4 the block method is applied for constructing approximate conformal mappings of some infinite domains with periodical structure.

Some references are made to my papers [56–70] concerning the block method, the contents of papers [59, 67] are partially reflected. Papers [71, 72] in which the use of the block method for Dirichlet's problem with nonanalytical boundary conditions is developed and the rapid approximate method of constructing Green's function of Laplace's operator on polygons is exposed, are not used in the book.

The prototype of the block method is the method of composite nets for solving boundary-value problems for Laplace's equation on domains with piecewise-smooth boundary (see [51–54]). In the method of composite nets, a polar net convergent at the vertex is constructed in order to increase accuracy, in the neighborhood of the vertices of the angles in which the derivatives of the required solutions have singularities. In the block method, by analogy, a block-sector on which the corresponding integral representation of the solution of the boundary-value problem is approximated with the account of the singularities of the derivatives of all orders is defined in the vicinity of each vertex. Paper [55] served for the transition from the method of composite nets to the block method. In [55], an approximate analytic representation of a solution via some quasipolynomials (functions of singularities) with unknown coefficients whose order is consistent with the step of the net on the rest part of the domain is used in the vicinity of the vertices of the angles.

In this book I speak only of the exponentially convergent block method and do not touch upon other approximate methods of solving boundary-value problems for Laplace's equation and constructing conformal mappings of simply-connected and multiply-connected domains. The following monographs are devoted to other approximate methods of solving these problems: M. A. Aleksidze [2], J.-P. Aubin [3]; K. I. Babenko [4]; I. Babuška, M. Práger and E. Vitásek [5]; N. S. Bakhvalov, N. P. Zhidkov and G. M. Kobel'kov [6]; I. S. Berezin and N. P. Zhidkov [7]; L. Collatz [9]; E. G. D'yakonov [10,11]; P. F. Fil'chakov [12]; V. P. Fil'chakova [13]; G. Forsythe and W. Wasow [14]; D. Gaier [15]; S. K. Godunov and V. S. Ryabenkiĭ [16]; L. V. Kantorovich and V. I. Krylov [19]; W. Koppenfels and F. Stallmann [22]; G. I. Marchuk and V. I. Agoškov [25]; G. I. Marchuk and Yu. A. Kuznetsov [26]; G. I. Marchuk and V. I. Lebedev [27]; G. I. Marchuk and V. V. Šaidurov [28]; S. G. Mikhlin [30, 31]; A. Mitchell and R. Wait [32]; L. A. Oganesyan and L. A. Rukhovets [33]; V. S. Ryabenkiĭ [36]; V. S. Ryabenkiĭ and A. F. Filippov [37]; A. A. Samarskiĭ [38]; A. A. Samarskiĭ and V. B. Andreev [39]; A. A. Samarskiĭ and A. V. Gulin [40]; A. A. Samarskiĭ, R. D. Lazarov, and V. L. Makarov [41]; A. A. Samarskiĭ and E. S. Nikolaev [42]; V. K. Saulyev [43]; G. Strang and G. Fix [46]; V. I. Vlasov [47].

The Author

Evgenii A. Volkov, Ph.D., D.Sc. (Phys.-Math.), is a Leading Researcher at Steklov Mathematical Institute of the Russian Academy of Sciences. He was born in 1926 in Tula, Russia. In 1943–1946 he was a Designer at a Munition Factory, Petropavlovsk, U.S.S.R. In 1946–1951 as a student of the Mechanical-Mathematics Department of Moscow State University he fulfilled a number of original investigations in the field of computing devices. Since 1951 he has been a Researcher at the Academy of Sciences, Moscow.

Dr. Volkov received his Ph.D. and D.Sc. degrees at Steklov Mathematical Institute of the U.S.S.R. Academy of Sciences in 1954 and 1967 respectively. From 1958 to 1970 as a Consultant at a research institute, Dr. Volkov took an active part in the realization of State Programs. In 1954–1958 he was an Associate Professor at the Moscow Physico-Technical Institute, and in 1971–1980 he was Professor at the Moscow Institute of Physical Engineering. Currently he is the Assistant Editor-in-Chief of the *Proceedings of the Steklov Mathematical Institute.*

Dr. Volkov fulfills scientific investigations in the field of applied mathematics (numerical methods for solving ordinary and partial differential equations, some nonlinear problems, etc.) from 1949, and possesses great experience in using high-speed computers from 1952. He published over 100 scientific articles in which he combined novelty, strict mathematical substantiation, and practical trends. Dr. Volkov is the author of the book *Numerical Methods*, Nauka Publ., Moscow, 1982, which has had two editions in Russian and was translated into English and Spanish.

Contents

Chapter 1

Approximate Block Method for Solving the Laplace Equation on Polygons

1. Setting up a Mixed Boundary-Value Problem for the Laplace Equation on a Polygon

1.1. Description of the Boundary of a Polygon

Before setting up a boundary-value problem we shall discuss the method of describing the boundary of a multiply-connected polygon.

Suppose a finite sequence of points $P_1, P_2, \ldots, P_\varkappa, \varkappa \geq 2$, is given in the complex plane z, $z = x + iy$. We set $P_0 = P_\varkappa$, $P_{\varkappa+1} = P_1$, i.e., understand P_0 to be the same point as $P_\varkappa$, and P_1 to be $P_{\varkappa+1}$. Figure 1 shows a sequence of points corresponding to $\varkappa = 5$. In what follows we shall not mark the points P_0 and $P_{\varkappa+1}$ in the figures.

It can be assumed that some points of the given sequence, $P_1, P_2, \ldots, P_\varkappa$, except for the points which have the neighboring numbers or the numbers 1 and $\varkappa$, coincide geometrically. Figure 2 shows a sequence of points in which P_2 and P_4 coincide.

Successively connecting the points $P_1, P_2, \ldots, P_\varkappa, P_{\varkappa+1}$ by line segments, we get a closed polygonal line Γ. We denote by γ_j the segment whose endpoints are P_j and P_{j+1} and call it a side of the polygonal line Γ. The points $P_1, P_2, \ldots, P_\varkappa$ are the vertices of the polygonal line Γ. In the general case different sides of the closed polygonal line

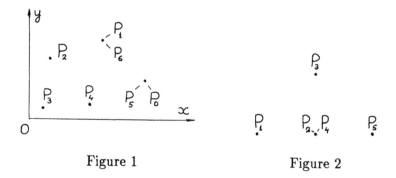

Figure 1   Figure 2

Γ, including the sides which have nonneighboring numbers, may intersect both at separate points and along segments of positive length.

Before considering the following definition, let us recall that a domain is an open connected set of points of the complex plane z.

Definition 1.1. *Let γ be a set of points of the complex plane z which belong to at least one side of the polygonal line Γ, i.e., $\gamma = \gamma_1 \cup \gamma_2 \cup \ldots \cup \gamma_\varkappa$. The closed polygonal line Γ is attainable from the inside (from the outside) if the following conditions are fulfilled:*

(a) there is a finite (infinite) domain D whose boundary coincides with γ;

(b) there is no segment of positive length lying at the intersection of more than two sides of the polygonal line Γ;

(c) if any two sides of the polygonal line Γ intersect along a segment of positive length, then every point of this segment, except, maybe, its endpoints, has a neighborhood entirely belonging to $D \cup \gamma$;

(d) the direction of traverse of the polygonal line Γ corresponding to the order of numbering of its vertices coincides with the generally accepted positive traverse of the boundary of the domain D for which the domain D is locally on the left of Γ.

Definition 1.2. *Closed polygonal lines attainable from the inside or from the outside are called admissible polygonal lines.*

We denote by $\alpha_j \pi$, $0 < \alpha_j \leq 2$, the magnitude of the angle with vertex P_j made by the sides γ_{j-1} and γ_j ($\gamma_0 = \gamma_\varkappa$) of the admissible closed polygonal line Γ reckoned counterclockwise from γ_j to γ_{j-1}

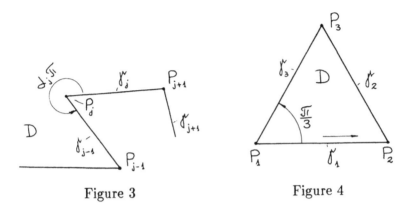

Figure 3   Figure 4

(see Fig. 3).

Here are some explanatory examples. Figures 4–7 show admissible polygonal lines, attainable from the inside, and Figures 8–11 show admissible polygonal lines attainable from the outside (verify conditions (a)–(d) in Definition 1.1). Arrows indicate the direction of positive traverse of some polygonal lines.

All the vertices of the closed polygonal line shown in Fig. 9 coincide with the points of the sequence presented in Fig. 2. The vertices of the polygonal lines shown in Figs. 4 and 8 are numbered in opposite directions. The first of them is attainable from the inside and its three angles have the magnitude $\pi/3$, i.e., $\alpha_1 = \alpha_2 = \alpha_3 = 1/3$. The other polygonal line is attainable from the outside and its angles are equal to $5\pi/3$, $\alpha_1 = \alpha_2 = \alpha_3 = 5/3$. However, by changing the direction of numbering of the vertices we cannot always obtain an admissible polygonal line with another orientation, i.e., a polygonal line attainable from the outside instead of that attainable from the inside or vice versa. For instance, when we renumber the vertices of the polygonal line shown in Fig. 6 or Fig. 7 in the reverse order, it can no longer be attained from the inside (condition (d) in Definition 1.1 is violated) and, evidently, cannot be attained from the outside since there is no infinite domain whose boundary is this polygonal line. It is also evident that when the vertices of the polygonal lines shown in Figs. 9 and 11 are renumbered in the reverse order, the polygonal

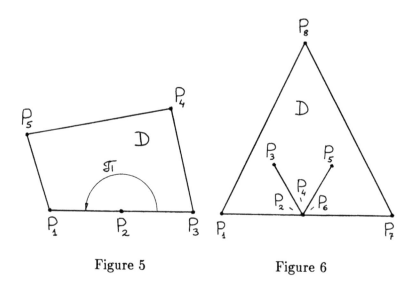

Figure 5   Figure 6

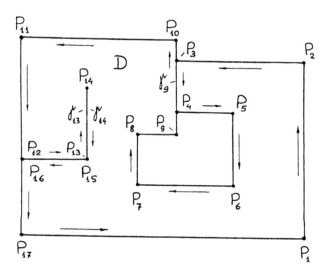

Figure 7

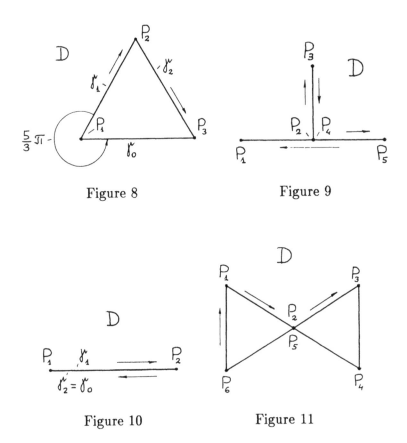

Figure 8   Figure 9

Figure 10   Figure 11

lines are no longer attainable from the outside and do not become polygonal lines attainable from the inside.

Figure 12 shows a closed polygonal line with four vertices and sides for which condition (b) in Definition 1.1 is violated since its four sides are superimposed.

The simplest polygonal line with only two vertices shown in Fig. 10 is attainable from the outside and, in particular, satisfies conditions (b) and (c) of Definition 1.1. Condition (c) has been introduced in Definition 1.1 to exclude closed polygonal lines one traverse of which entails two traverses of the boundary of the domain D.

More than two vertices (but not sides) of admissible polygonal lines may overlap. In particular, the polygonal line attainable from the inside, shown in Fig. 6, has three coincident geometric vertices. Some vertices of an admissible polygonal line may coincide geometrically with the interior points of some of its sides. For instance, in Fig. 7 the vertices P_3 and P_4 are within the side γ_9. The polygonal line shown in Fig. 5 has an angle of magnitude π with a vertex P_2. In some cases the introduction of angles of magnitude π has a definite sense which we will elucidate later.

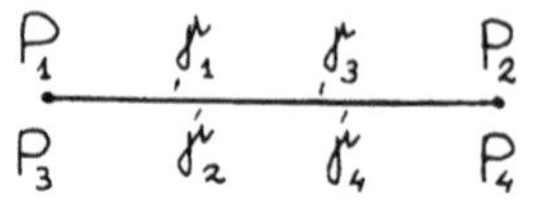

Figure 12

Definition 1.3. *An open finitely-connected bounded domain Ω in the complex plane z is a polygon if its boundary can be composed of a finite number of pairwise nonintersecting admissible closed polygonal lines one of which is attainable from the inside (the external boundary of the domain Ω), and the others are attainable from the outside.*

In the case when the boundary of the polygon Ω serves as only one polygonal line attainable from the inside, the polygon is simply-connected.

The vertices, sides and angles of polygonal lines that form the boundary of the polygon Ω are the vertices, sides and angles of the polygon Ω itself. For the sake of convenience, we introduce the following unbroken numeration of the vertices and sides of the multiply-connected polygon Ω. We preserve the numbers of the vertices and sides of the polygonal line attainable from the inside, i.e., the external boundary of the polygon Ω. Then we continue to number the vertices and, respectively, the sides of a certain closed polygonal line, attainable from the outside, constituting a part of the boundary, by natural numbers. The new numeration is in the same order as the original numeration of the vertices and sides of the polygonal line. Then, by analogy, we continue to number the vertices and sides of the next polygonal line, attainable from the outside, constituting the boundary of the polygon Ω, and so on. The total number of vertices and sides of the polygon Ω will be denoted by N.

Figure 13 shows a doubly-connected polygon with $N = 10$ vertices

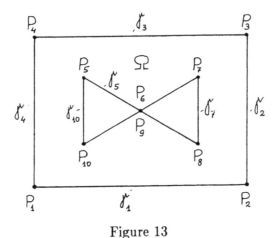

Figure 13

and sides and Figs. 4–7 show simply-connected polygons with $N = 3, 5, 8$ and 17 vertices and sides, respectively. It is useful to remember that every side of the polygon Ω has a number coincident with the number of the vertex which is at the end of the side from which the traverse begins and is continued in the direction of the numeration of the vertices on the corresponding closed polygonal line (see Fig. 13).

1.2.   Setting up a Boundary-Value Problem

Let us consider on a, generally speaking, multiply-connected polygon Ω a mixed boundary-value problem

$$\Delta u = 0 \quad \text{on} \quad \Omega, \tag{1.1}$$

$$\nu_j u + \bar{\nu}_j u'_n = \varphi_j(s) \quad \text{on} \quad \gamma_j, \qquad j = 1, 2, \ldots, N, \tag{1.2}$$

where $\Delta \equiv \partial^2/\partial x^2 + \partial^2/\partial y^2$ is Laplace's operator, γ_j is a side of the polygon Ω, ν_j is a parameter equal to 0 or 1, $\bar{\nu}_j = 1 - \nu_j$, u'_n is a derivative in the direction of the inner normal to γ_j, $\varphi_j(s)$ is a specified algebraic polynomial of the arc length s reckoned along γ_j, $1 \leq \nu_1 + \nu_2 + \ldots + \nu_N \leq N$.

The defined parameter ν_j specifies the type of the boundary condition on γ_j. If $\nu_j = 1$ and $\bar{\nu}_j = 1 - \nu_j = 0$, then, in accordance with

(1.2), a boundary condition of the first kind is defined on the side γ_j. If $\nu_j = 0$ and, consequently, $\bar{\nu}_j = 1 - \nu_j = 1$, then a boundary condition of the second kind is defined on γ_j. The value of the parameter ν_j is independently specified on every side γ_j with the only requirement that a boundary condition of the first kind be at least on one side. In a special case, when $\nu_1 = \nu_2 = \ldots = \nu_N = 1$, problem (1.1), (1.2) is the Dirichlet problem. Another special case, $\nu_1 = \nu_2 = \ldots = \nu_N = 0$, corresponds to the Neumann problem which is considered separatly in Sec. 9 on a simply-connected polygon. Paper [59] is devoted to the Neumann problem on multiply-connected polygons.

Let us elucidate the concept of the derivative u'_n in the direction of the inner normal to the boundary of the polygon Ω. We note first of all that the normal is defined at all points of the sides of the polygon, except, maybe, their ends lying at the vertices. At the vertices of the polygon there is no normal to its boundary, except for the unique case when two adjacent sides form an angle equal to π. Therefore, when we speak of the derivative of the solution along the inner normal to the side γ_j and, in particular, under condition (1.2), we mean that the endpoints of γ_j, i.e., corresponding vertices of the polygon Ω, are not considered. It remains to determine the direction of the inner normal. If the sides of a polygon intersect only at their endpoints and are not superimposed as, say, in Figs. 4, 5 and 13, then the direction of the inner normal to the sides is easy to find. It corresponds to the only orthogonal direction into the interior of the domain. When there are overlapping sides as, for instance, in Fig. 7, then the direction of the inner normal has to be additionally determined. In Fig. 7 the innner normal to the sides γ_3, γ_5, γ_{14} (the vertices P_j and P_{j+1} are the endpoints of γ_j) is directed to the right and that to the sides γ_1, γ_9, γ_{13} goes in the opposite direction, i.e., to the left. An inner normal always lies on the left of every side when a closed polygonal line (to which the side belongs) is traversed in the direction of numeration of the vertices, i.e., in the positive direction.

Let us now pay attention to an important fact that no consistency conditions at the vertices P_j are imposed on the polynomials given in the boundary conditions (1.2). In particular, for $\nu_{j-1} = \nu_j = 1$ the values φ_{j-1} and φ_j at $P_j \in \gamma_{j-1} \cap \gamma_j$ may not coincide, i.e., discontinuities of the first kind of the defined boundary values are permissible at the vertices. For $\nu_{j-1} = \nu_j = 0$ and $\alpha_j = 1$ the values φ_{j-1} and φ_j at P_j may not coincide either, i.e., the value of the given normal derivative may have a discontinuity of the first kind at the point P_j which is the vertex of an angle equal to π. Finally, for $\nu_{j-1} + \nu_j = 1$ and $\alpha_j = 1/2$ or $\alpha_j = 3/2$ (P_j is the vertex of the angle $\pi/2$ or $3\pi/2$) the limiting value at the vertex P_j of the given normal derivative on one of the sides, γ_{j-1} or γ_j, may not coincide with the corresponding derivative, at P_j, of the boundary values given on the other side.

Some domains, as, for instance, those shown in Figs. 6 and 7, may have the so-called cuts. The domain shown in Fig. 6 has been obtained from a triangle (with vertices P_1, P_7, P_8) with two rectilinear oblique cuts made from the middle of the lower side. Two sides of the polygonal line which constitute the boundary of the resulting polygon lie on each cut. On two sides lying on the cut the boundary conditions in the corresponding problem (1.1), (1.2) are also independent of each other. For instance, they may both be conditions of the first kind, and, say, on one side (on one edge of the cut) the given boundary values are equal to unity and on the other side (on the other edge of the cut) they are zero. On these two sides (the two edges of the cut) the boundary conditions may be of different kinds. The domain shown in Fig. 7 has a cut with a break on which four sides γ_{12}, γ_{13}, γ_{14} and γ_{15} lie. On each side the boundary conditions are independent. Note that the endpoints of the cuts are the vertices of the angles of the polygon equal to 2π.

Definition 1.4. Let $\Omega_j \subset \Omega$ be an arbitrary convex subdomain on whose boundary lies the side γ_j of the polygon Ω and the inner normal (introduced earlier) to γ_j is directed into Ω_j. We say that the function f, defined on the open polygon Ω, is continuous up to the boundary of the polygon Ω, except a set Ξ of its vertices, if this function is continuous on Ω and can be extended from Ω to every side γ_j, $1 \leq j \leq N$, so that the extended function is continuous on $\Omega_j \cup \gamma_j \setminus \Xi_j$, where $\Xi_j = \Xi \cap (P_j \cup P_{j+1})$.

Remark 1.1. Let f be a function subject to Definition 1.4. It is evident that its values extended from Ω (by continuity) to $\gamma_j \setminus \Xi_j$ are uniquely defined. We say that the function f assumes these values on $\gamma_j \setminus \Xi_j$. It stands to reason that the values the function f assumes (by continuity, from the side of the inner normal) may not coincide on two superimposed sides, say, those lying on the cut. On one edge of the cut, except for its end, the function f may assume values, say, equal to unity and on the other edge, equal to zero. The function $f = (\arg z)/2\pi$ may serve as an example when the polygon Ω has a cut with its end at the point $x = 0$ $(z = 0)$ lying on the real semiaxis $x \geq 0$.

Let Π be the set of all vertices of the polygon Ω for which the following conditions are fulfilled:

(a) at a vertex lie the endpoints of two adjacent sides on which boundary conditions of the first kind are simultaneously specified,

(b) the boundary values defined on these sides do not coincide at the vertex.

Definition 1.5. *The solution of the boundary-value problem* (1.1), (1.2) *is a function u with the following properties:*

(1) u is bounded and satisfies Laplace's equation on the polygon Ω,

(2) u is continuous up to the boundary of the polygon Ω, except for the set Π of its vertices at which the specified boundary values are discontinuous,

(3) on the polygon Ω the function u has partial derivatives $\partial u/\partial x$ and $\partial u/\partial y$, continuous up to its boundary, except for some of its vertices,

(4) for $\nu_j = 1$ the function u assumes values coinciding with the values of the given polynomial $\varphi_j(s)$ at the points $\gamma_j \setminus \Pi$,

(5) for $\nu_j = 0$ the derivative of the function u along the inner normal to γ_j assumes values equal to those of the given polynomial $\varphi_j(s)$ at all points of the side γ_j, except, maybe, one or two of its ends.

Theorem 1.1. *There exists a unique solution u of the boundary-value problem* (1.1), (1.2) *in the sense of Definition 1.5.*

The proof of Theorem 1.1 is beyond the scope of this book. The existence of a solution is established for the case of the simply-connected polygon Ω by the methods of the theory of functions of

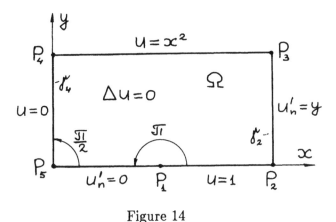

Figure 14

a complex variable and the Keldysh–Sedov formula (see [23]) and, for the general case, by the variational method (see [45], [29]) or else, by Schwarz's alternating method. In the case of Dirichlet's problem, the existence can also be proved by the classical method of subharmonic functions and barriers (see [20], [23]). The uniqueness of a bounded solution of problem (1.1), (1.2) follows from the maximum principle.

The following example of a boundary-value problem of type (1.1), (1.2) may serve as an illustration. In Fig. 14 we depict what in our sense is a pentagon with vertices P_1, P_2, P_3, P_4, P_5 having complex coordinates $z_1 = 1$, $z_2 = 2$, $z_3 = 2 + i$, $z_4 = i$, $z_5 = 0$, respectively. The boundary conditions are written out next to each side. The point P_1 is the vertex of an angle equal to π ($\alpha_1 = 1$). The other angles of the polygon are equal to $\pi/2$.

The typical feature of problem (1.1), (1.2) is that the type of the boundary condition is preserved on each side of the polygon Ω and the right-hand side of the boundary condition on the whole side is defined by a single algebraic polynomial of the arc length of the boundary. Therefore, if, in some problem, there are points of the rectilinear part of the boundary at which either the type of the boundary condition varies or the right-hand side of the boundary condition varies by a jump (is piecewise polynomial), then we can introduce vertices of angles, equal to π, at the indicated points and in this way reduce the problem to form (1.1), (1.2). The arrangement of conditions on the sides of the polygon Ω in problem (1.1), (1.2) is arbitrary under the only assumption that a condition of the first kind is defined at least on one side. In particular, in the problem presented in Fig. 14, $\nu_1 = \nu_3 = \nu_4 = 1$ (conditions of the first kind) and $\nu_2 = \nu_5 = 0$ (conditions of the second kind).

Remark 1.2. The solution u of the boundary-value problem (1.1), (1.2) is not only continuous and continuously differentiable up to the boundary of the polygon Ω (except for a certain set of its vertices), but also admits of an analytical (harmonic) continuation across every side γ_j, except, maybe, its ends since the right-hand sides of the boundary conditions are analytical (see Remark 6.3 below). However, the derivatives of the solution u usually possess singularities at the vertices of the polygon Ω even if the functions $\varphi_j(s)$ are linear and, in particular, constants. If the polygon Ω has a reentrant angle, as, for instance, the angle with vertex P_j in Fig. 3, then, as a rule, even the first derivatives of the solution are unbounded in the vicinity of the vertex even when the boundary conditions of the first or second kind (in any combinations) on its sides are homogeneous. The construction of the solution of problem (1.1), (1.2) is studied in detail in paper [50] in which, in particular, the necessary and sufficient conditions are established for the specified smoothness of the solution on the simply-connected polygon Ω in terms of the classes[1] $C_{k,\lambda}(\Omega)$, $0 \le k < \infty$, $0 < \lambda < 1$. These conditions include a number of local linear conditions of consistency at the vertices of the angles for the functions $\varphi_j(s)$ and global requirements expressed in terms of the integrals (with unknown kernels) of the indicated functions along the boundary. In the general case, these conditions cannot be verified in practice and, as a rule, are not satisfied. Since the smoothness of the required solution is unessential up to the vertices of the polygon Ω for the approximate block method of solving the boundary-value problem (1.1), (1.2) discussed below, we do not investigate here the behavior of the solution in the vicinity of the vertices but refer the

---

[1]The function $f \in C_{k,\lambda}(\Omega)$, $k \ge 0$, $\lambda > 0$, if f is k times differentiable on Ω and all its partial derivatives of order k satisfy Hölder's condition with exponent λ on Ω.

reader to papers [49, 50].

2. A Finite Covering of a Polygon by Blocks of Three Types

The first step in seeking an approximate solution of the boundary-value problem (1.1), (1.2) by the method being considered is the construction of a covering of the polygon Ω by a finite number of overlapping sectors of disks, half-disks and disks which we shall call blocks. In Chapters 3 and 4 we shall show how some other blocks can be used. Here we only point out that an approximate solution of problem (1.1), (1.2) can be found on every block as a certain harmonic elementary function.

Suppose we are given, generally speaking, a multiply-connected polygon Ω with vertices P_j, $j = 1, 2, \ldots, N$. We choose a finite number of points P_q, $q = N + 1, N + 2, \ldots, L$, lying inside (i.e., not at the endpoints) of some sides of the polygon Ω and also several points P_m, $m = L + 1, L + 2, \ldots, M$, lying strictly within the polygon Ω, $N \leq L \leq M$. There may be several points chosen on some sides of the polygon and none on the other sides. Later we shall explain the aim and means of choosing the indicated points.

We denote by $j(q)$ the number of the side of the polygon Ω on which the point P_q lies, $N < q \leq L$. Suppose that l_j, $1 \leq j \leq N$, is a ray emanating from the vertex P_j and directed along the side γ_j, l_q, $N < q \leq L$, is a ray emanating from the point P_q and passing along $\gamma_{j(q)}$ in the direction opposite to the vertex $P_{j(q)}$, and l_m, $L < m \leq M$, is a ray which emanates from the point P_m and whose fixed direction is arbitrary. We introduce a polar coordinate system r_μ, θ_μ with pole at P_μ and angle θ_μ reckoned from the ray l_μ counterclockwise, $\mu = 1, 2, \ldots, M$.

Recall that the interior angle of the polygon Ω, whose vertex is at P_j, has a magnitude $\alpha_j \pi$, $0 < \alpha_j \leq 2$, $1 \leq j \leq N$. We set $\alpha_p = 1$, $N < p \leq L$, and introduce the notations

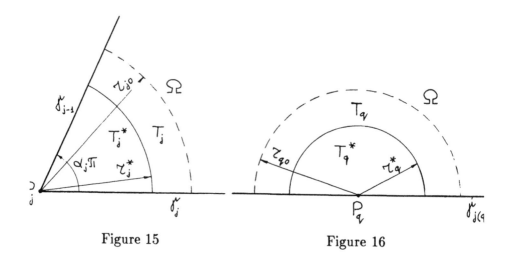

Figure 15   Figure 16

$$T_q(r) = \{(r_q, \theta_q) : 0 < r_q < r, 0 < \theta_q < \alpha_q \pi\}, \qquad (2.1)$$

$1 \leq q \leq L$, is a sector of a disk (a half-disk for $\alpha_q = 1$),

$$T_m(r) = \{(r_m, \theta_m) : 0 \leq r_m < r, 0 \leq \theta_m < 2\pi\}, \qquad (2.2)$$

$L < m \leq M$, is a disk.

We specify numbers $r_{\mu 0}$, r_μ^* satisfying the inequalities $0 < r_\mu^* < r_{\mu 0}$, $1 \leq \mu \leq M$, and, for brevity, introduce notations

$$T_\mu = T_\mu(r_{\mu 0}), \qquad T_\mu^* = T_\mu(r_\mu^*). \qquad (2.3)$$

The blocks (sectors of disks, half-disks, disks) T_μ^* are called *basic blocks* and the blocks T_μ are called *extended blocks*, $1 \leq \mu \leq M$. Every basic block T_μ^* is a part of the corresponding extended block T_μ. Figures 15–17 show blocks of all three types being considered. The sides of the polygon Ω and curvilinear parts of the boundaries of the basic blocks are shown by solid lines and the curvilinear parts of the boundaries of the extended blocks are shown by dash lines.

We assume that the additional points P_q, $N < q \leq L$, on the sides and the points P_m, $L < m \leq M$, inside the polygon Ω as well as the numbers $r_{\mu 0}, r_\mu^*, 1 \leq \mu \leq M$, are chosen such that the following four conditions are fulfilled.

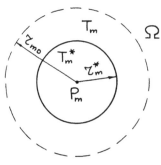

Figure 17

I. $T_\mu \subset \Omega$, $1 \leq \mu \leq M$, i.e., extended open blocks lie entirely in the open polygon Ω.

II. For $1 \leq j \leq N$ the number r_{j0} is smaller than the length of the minimal of the two sides of the polygon Ω emanating from the vertex P_j, and for $N < q \leq L$ the number r_{q0} is smaller than the distance from the point P_q to the nearest end of the side $\gamma_{j(q)}$ on which this point lies.

III. For $1 \leq \mu \leq M$ the curvilinear part of the boundary of the extended block T_μ does not touch the vertices of the polygon Ω but may touch the sides at their interior points.

IV. $\overline{T}_1^* \cup \overline{T}_2^* \cup \ldots \cup \overline{T}_M^* = \overline{\Omega}$, i.e., closed basic blocks of "smaller dimensions" form (with certain intersections) a covering of the closed polygon $\overline{\Omega}$ and, all the more so, the closed extended blocks form the indicated covering.

Remark 2.1. The double covering of a polygon by basic and extended blocks is necessary to find an approximate solution of the boundary-value problem (1.1), (1.2) by the method being discussed (see Remark 6.4 below). In what follows, when speaking of a covering of the polygon Ω, we shall first of all bear in mind the existence of a covering by closed basic blocks considering conditions I–III to be satisfied for extended blocks. We shall obtain an approximate solution of problem (1.1), (1.2) on closed basic blocks.

In practice, the covering of the polygon Ω satisfying the formulated requirements I–IV, is constructed as follows. We first define the sectors T_j and T_j^*, $1 \leq j \leq N$, with vertices lying at the corre-

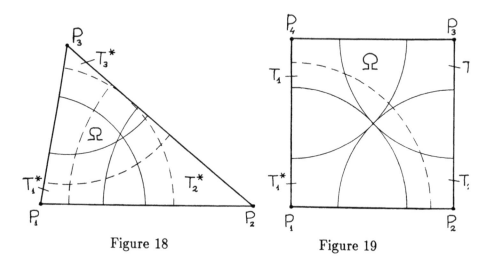

Figure 18   Figure 19

sponding vertex P_j of the polygon Ω and the lateral sides lying on
the sides of the polygon Ω emanating from P_j. We choose r_{j0}, i.e.,
the radius of the disk from which the extended sector T_j is cut, in
accordance with requirements I–III and take the radius r_j^* of the disk
from which the basic sector T_j^* is cut somewhat smaller than r_{j0}. For
triangles and certain quadrilaterals and pentagons these sectors may
form the required covering as, for instance, those shown in Figs. 18
and 19. In the general case, the sectors T_j^*, $1 \le j \le N$, are insufficient
for obtaining the required covering. Therefore, we have to construct
half-disks T_q^*, $N < q \le L$, next to some sides of the polygon and, in
general, construct disks T_m^*, $L < m \le M$, in the interior of the polygon Ω in order to satisfy requirement IV and fulfill conditions I–III
for the corresponding extended blocks T_q and T_m. For instance, in
the case when Ω is a rectangle with ratio of sides ≥ 2, it is necessary
to construct half-disks next to the long sides of the rectangle in order
to obtain a covering and when all sides of a polygon are smaller than
the radius of the maximum disk it includes, the covering necessarily
contains disks.

Let us consider some more examples on the construction of cov-
erings of polygons. Figure 20 shows a covering corresponding to a
boundary-value problem on the pentagon presented in Fig. 14. We

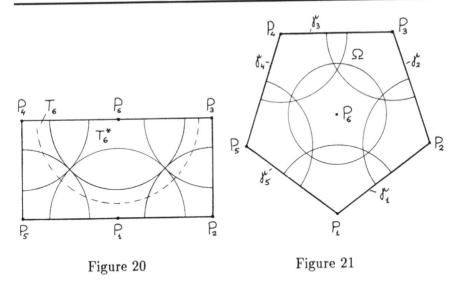

Figure 20   Figure 21

had to introduce an additional point P_6 at the middle of the side γ_3 and, respectively, block-half-disks T_6^* and T_6. In this case $N = 5$, $L = 6$, $M = L$. To obtain a covering for a regular pentagon (Fig. 21), in addition to the sectors $T_1^*, T_2^*, \ldots, T_5^*$, we also used a disk T_6^* with center at P_6 lying at the center of the pentagon, i.e., inside Ω ($N = 5$, $L = N$, $M = 6$). When we defined a covering of a hexagonal L-shaped domain with vertices $P_1, P_2, \ldots, P_6$ (Fig. 22), we used additional half-disks T_7^*, T_8^* and disks T_9^*, T_{10}^* ($N = 6$, $L = 8$, $M = 10$). Finally, Fig. 23 shows a covering of a doubly-connected polygon (P_1, P_2, P_3, P_4 are vertices of its external boundary) with an inner cut whose vertices (ends) are points P_5 and P_6. Besides the basic sectors $T_1^*, T_2^*, \ldots, T_6^*$, we also use additional half-disks T_7^* and T_8^* ($N = 6$, $L = 8$, $M = L$).

For the sake of simplicity, we do not show the boundaries of the extended blocks T_μ (by dash lines) in Figs. 21–23. Condition IV for the basic blocks T_μ^* is fulfilled in these figures and it is clear that conditions I–III can be fulfilled for the extended blocks T_μ for the chosen T_μ^*.

It is not difficult to construct a covering for an arbitrary finitely-connected polygon Ω by sectors, half-disks and disks which would satisfy requirements I–IV, although a large number of blocks may be

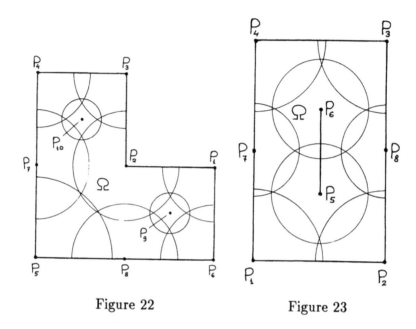

Figure 22   Figure 23

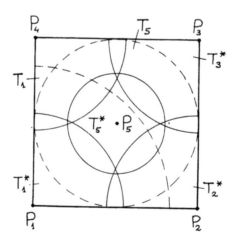

Figure 24

needed. A choice of covering is, evidently, not unique. For instance, besides the covering of the square shown in Fig. 19, the covering depicted in Fig. 24 is also permissible, with the use of additional block-disks T_5^* and T_5 with centers at the center of the square.

In Sec. 1–3 we assume that the covering of the polygon Ω satisfying requirements I–IV is defined.

3. Representation of the Solution of a Boundary-Value Problem on Blocks

Here we construct integral representation of the solution of the boundary-value problem (1.1), (1.2) on extended blocks, namely, sectors and half-disks, except for the curvilinear parts of the boundaries of the blocks. The resulting integral representations are analogs of the well-known Poisson integral for a circle. The solution is given on block-sectors and block-half-disks in terms of the integrals (with elementary kernels) of the values of the solution itself on the curvilinear part of the boundary of the block and also in terms of the specified boundary values of the required solution or of its normal derivative along the sides of the polygon Ω to which the corresponding block is adjacent. These representations lie at the basis of the construction of an approximate block method for solving the boundary-value problem (1.1), (1.2) which is carried out in Sec. 4 and 5.

3.1. Carrier Functions

After constructing a finite covering of the polygon Ω by blocks in accordance with the rules formulated in Sec. 2, we define, on extended block-sectors and half-disks, elementary harmonic functions which satisfy the same boundary conditions as the required solution of problem (1.1), (1.2) at the intersection of the rectilinear parts of the boundaries of these blocks with the boundary of the polygon Ω. These functions explicitly appear in the representations of the solution of the boundary-value problem on blocks considered below. We call them carrier functions. They contain (carry) information concerning the specified boundary conditions on the indicated sides of the polygon Ω.

Suppose that P_j is a vertex of the polygon Ω on which the bound-ary-value problem (1.1), (1.2) is posed and T_j^*, T_j are the correspond-ing (basic and extended) block-sectors entering into the covering of the polygon Ω. The sides of the sectors T_j^* and T_j lie on the sides γ_{j-1} and γ_j of the polygon Ω making an interior angle of magnitude $\alpha_j\pi$, $0 < \alpha_j \le 2$. Here and in what follows, for the sake of convenience, we understand γ_{j-1} to be the side preceding the side γ_j in the positive traverse of the polygonal line to which the vertex P_j belongs. This side either is actually labeled by $j - 1$ or its actual number is the largest on this polygonal line. Correspondingly, we understand φ_{j-1} and ν_{j-1} to be a polynomial and a parameter defined on γ_{j-1} in the boundary condition (1.2).

Let r_μ, θ_μ be a polar coordinate system with pole P_μ introduced in Sec. 2, $1 \le \mu \le M$.

We represent the polynomials φ_{j-1} and φ_j defined in the bound-ary conditions (1.2) on the sides γ_{j-1} and γ_j as

$$\varphi_{j-1} = \sum_{k=0}^{m_{j-1}} a_{jk} r_j^k, \qquad \varphi_j = \sum_{k=0}^{m_j} b_{jk} r_j^k, \tag{3.1}$$

where a_{jk}, b_{jk} are numerical coefficients and m_{j-1}, m_j are the degrees of these polynomials.

On the closed extended block-sectors $\overline{T}_j$, $1 \le j \le N$, we de-fine a carrier function $Q_j(r_j, \theta_j)$ as it depends on the values of the parameters ν_{j-1} and ν_j as follows:

(1) for $\nu_{j-1} = \nu_j = 1$ (boundary conditions of the the first kind are defined on γ_{j-1} and γ_j)

$$Q_j(r_j, \theta_j) = b_{j0} + \frac{a_{j0} - b_{j0}}{\alpha_j\pi}\theta_j$$

$$+ \sum_{k=1}^{m_{j-1}} a_{jk}\xi_{jk}(r_j, \theta_j) + \sum_{k=1}^{m_j} b_{jk}\xi_{jk}(r_j, \alpha_j\pi - \theta_j), \tag{3.2}$$

where

$$\xi_{jk}(r_j, \theta_j) =$$

$$= \begin{cases} r_j^k \dfrac{\theta_j \cos k\theta_j + \ln r_j \sin k\theta_j}{\alpha_j \pi \cos k\alpha_j \pi}, & \sin k\alpha_j \pi = 0, \\[3mm] r_j^k \dfrac{\sin k\theta_j}{\sin k\alpha_j \pi}, & |\sin k\alpha_j \pi| \geq \tfrac{1}{2}, \\[3mm] \dfrac{r_j^k \sin k\theta_j - \sigma_{jk} r_j^{k'} \sin k'\theta_j}{\sin k\alpha_j \pi}, & 0 < |\sin k\alpha_j \pi| < \tfrac{1}{2}, \end{cases} \tag{3.3}$$

$k' = [k\alpha_j + 1/2]/\alpha_j$, [ ] is the sign of the integer part, $\sigma_{jk} = (6r_{j0}/5)^{k-k'}$, $\alpha_j \pi$ is the magnitude of the vertex angle of the sector T_j,

(2) for $\nu_{j-1} = \nu_j = 0$ (boundary conditions of the second kind are defined on γ_{j-1} and γ_j)

$$Q_j(r_j, \theta_j) = \sum_{k=0}^{m_{j-1}} a_{jk} \eta_{jk}(r_j, \theta_j) + \sum_{k=0}^{m_j} b_{jk} \eta_{jk}(r_j, \alpha_j \pi - \theta_j), \tag{3.4}$$

where

$$\eta_{jk}(r_j, \theta_j) =$$

$$= \begin{cases} r_j^{k+1} \dfrac{\ln r_j \cos(k+1)\theta_j - \theta_j \sin(k+1)\theta_j}{(k+1)\alpha_j \pi \cos(k+1)\alpha_j \pi}, & \sin(k+1)\alpha_j \pi = 0, \\[3mm] r_j^{k+1} \dfrac{\cos(k+1)\theta_j}{(k+1)\sin(k+1)\alpha_j \pi}, & |\sin(k+1)\alpha_j \pi| \geq \dfrac{1}{2}, \\[3mm] \dfrac{r_j^{k+1} \cos(k+1)\theta_j - \sigma_{jk} r_j^{k'} \cos k'\theta_j}{(k+1)\sin(k+1)\alpha_j \pi}, & 0 < |\sin(k+1)\alpha_j \pi| < \dfrac{1}{2}, \end{cases} \tag{3.5}$$

$$k' = [(k+1)\alpha_j + 1/2]/\alpha_j, \qquad \sigma_{jk} = (6r_{j0}/5)^{k+1-k'},$$

(3) for $\nu_{j-1} = 0$, $\nu_j = 1$ (boundary conditions of the second kind and the first kind are defined on γ_{j-1} and γ_j, respectively)

$$Q_j(r_j, \theta_j) = \sum_{k=0}^{m_{j-1}} a_{jk} \vartheta_{jk}(r_j, \theta_j) + \sum_{k=0}^{m_j} b_{jk} \varkappa_{jk}(r_j, \alpha_j \pi - \theta_j), \tag{3.6}$$

where

$$\vartheta_{jk}(r_j, \theta_j) =$$

$$= \begin{cases} r_j^{k+1} \dfrac{\theta_j \cos(k+1)\theta_j + \ln r_j \sin(k+1)\theta_j}{(k+1)\alpha_j \pi \sin(k+1)\alpha_j \pi}, & \cos(k+1)\alpha_j \pi = 0, \\[3mm] -r_j^{k+1} \dfrac{\sin(k+1)\theta_j}{(k+1)\cos(k+1)\alpha_j \pi}, & |\cos(k+1)\alpha_j \pi| \geq \dfrac{1}{2}, \\[3mm] \dfrac{\sigma_{jk} r_j^{k'} \sin k'\theta_j - r_j^{k+1} \sin(k+1)\theta_j}{(k+1)\cos(k+1)\alpha_j \pi}, & 0 < |\cos(k+1)\alpha_j \pi| < \dfrac{1}{2}, \end{cases}$$ (3.7)

$$k' = (2[(k+1)\alpha_j] + 1)/2\alpha_j, \qquad \sigma_{jk} = (6r_{j0}/5)^{k+1-k'},$$

$$\varkappa_{jk}(r_j, \theta_j) =$$

$$= \begin{cases} r_j^{k} \dfrac{\theta_j \sin k\theta_j - \ln r_j \cos k\theta_j}{\alpha_j \pi \sin k\alpha_j \pi}, & \cos k\alpha_j \pi = 0, \\[3mm] r_j^{k} \dfrac{\cos k\theta_j}{\cos k\alpha_j \pi}, & |\cos k\alpha_j \pi| \geq \dfrac{1}{2}, \\[3mm] \dfrac{r_j^{k} \cos k\theta_j - \tau_{jk} r_j^{k''} \cos k''\theta_j}{\cos k\alpha_j \pi}, & 0 < |\cos k\alpha_j \pi| < \dfrac{1}{2}, \end{cases}$$ (3.8)

$$k'' = (2[k\alpha_j] + 1)/2\alpha_j, \qquad \tau_{jk} = (6r_{j0}/5)^{k-k''},$$

(4) for $\nu_{j-1} = 1$, $\nu_j = 0$ (boundary conditions of the first and the second kind are defined on γ_{j-1} and γ_j, respectively)

$$Q_j(r_j, \theta_j) = \sum_{k=0}^{m_{j-1}} a_{jk} \varkappa_{jk}(r_j, \theta_j) + \sum_{k=0}^{m_j} b_{jk} \vartheta_{jk}(r_j, \alpha_j \pi - \theta_j).$$ (3.9)

Remark 3.1. Direct verifications show that the function $Q_j(r_j, \theta_j)$ is harmonic and bounded on the open extended sector T_j, continuous on $\overline{T}_j$ everywhere, except for the point P_j (the vertex of the sector) for $\nu_{j-1} = \nu_j = 1$ and $a_{j0} \neq b_{j0}$ (i.e., for boundary conditions discontinuous at P_j), continuously differentiable on $\overline{T}_j \setminus P_j$ and satisfying the boundary conditions in accordance with items (4) and (5) of Definition 1.5, on $\gamma_{j-1} \cap \overline{T}_j$ and $\gamma_j \cap \overline{T}_j$, $1 \leq j \leq N$.

Remark 3.2. For the case of $\nu_{j-1} = \nu_j = 1$, we formally set the value of the carrier function (3.2) and the solution u of problem

(1.1), (1.2) at the vertex P_j equal to $(a_{j0} + b_{j0})/2$. This is convenient and does not contradict the case of $a_{j0} = b_{j0}$ when the solution u and function (3.2) are continuous at this vertex. For $\nu_{j-1} + \nu_j \leq 1$, the solution u of problem (1.1), (1.2) and the corresponding carrier function (3.4), (3.6) or (3.9) are always continuous at the vertex P_j.

Suppose (as in Sec. 2) that $j(q)$ is the number of the side of the polygon Ω which contains the point P_q, $N < q \leq L$, and to which the half-disk T_q, belonging to the covering of the polygon Ω is adjacent, s_q is the arc length s of the boundary of the polygon Ω corresponding to the point P_q (s increases in the direction of the positive traverse of the boundary). Let us represent the polynomial $\varphi_{j(q)}$ defined on the side $\gamma_{j(q)}$ as

$$\varphi_{j(q)} = \sum_{k=0}^{m_{j(q)}} c_{qk}(s - s_q)^k,$$

where c_{qk} are numerical coefficients. Let us define the carrier function

$$Q_q(r_q, \theta_q) = \sum_{k=0}^{m_{j(q)}} c_{qk} r_q^k \left(\nu_{j(q)} \cos k\theta_q + \bar{\nu}_{j(q)} r_q \frac{\sin(k+1)\theta_q}{k+1} \right) \quad (3.10)$$

for $N < q \leq L$.

Remark 3.3. It is evident that the function $Q_q(r_q, \theta_q)$ is a harmonic polynomial satisfying the given boundary condition on $\gamma_{j(q)}$, $N < q \leq L$. In particular, this harmonic polynomial is defined on the closed extended half-disk $\bar{T}_q$. For the sake of generality, we shall also call it a carrier function.

Finally, let

$$Q_p(r_p, \theta_p) = 0, \qquad L < p \leq M. \quad (3.11)$$

We set

$$R(m, m, r, \theta, \eta) = R(r, \theta, \eta) + (-1)^m R(r, \theta, -\eta), \quad (3.12)$$

$$R(1 - m, m, r, \theta, \eta) = R(m, m, r, \theta, \eta)$$
$$- (-1)^m R(m, m, r, \theta, \pi - \eta), \quad (3.13)$$

where $m = 0, 1$,

$$R(r, \theta, \eta) = \frac{1 - r^2}{2\pi(1 - 2r\cos(\theta - \eta) + r^2)} \tag{3.14}$$

is the kernel of the Poisson integral for a unit circle.

We specify the kernels

$$R_p(r_p, \theta_p, \eta) = R\left(\frac{r_p}{r_{p0}}, \theta_p, \eta\right), \qquad L < p \le M, \tag{3.15}$$

$$R_q(r_q, \theta_q, \eta) = R\left(\nu_{j(q)}, \nu_{j(q)}, \frac{r_q}{r_{q0}}, \theta_q, \eta\right), \qquad N < q \le L, \tag{3.16}$$

$$R_j(r_j, \theta_j, \eta) = \lambda_j R\left(\nu_{j-1}, \nu_j, \left(\frac{r_j}{r_{j0}}\right)^{\lambda_j}, \lambda_j \theta_j, \lambda_j \eta\right), \tag{3.17}$$

$$1 \le j \le N,$$

where $j(q)$ is the number of the side on which the point P_q lies, ν_{j-1}, ν_j, $\nu_{j(q)}$ are parameters entering into the boundary condition (1.2),

$$\lambda_j = 1/(2 - \nu_{j-1}\nu_j - \bar{\nu}_{j-1}\bar{\nu}_j)\alpha_j. \tag{3.18}$$

Remark 3.4. On rectilinear parts of the boundary of the extended block-sectors T_j and half-disks T_q corresponding to the kernels R_j, $1 \le j \le N$, R_q, $N < q \le L$, these kernels satisfy the homogeneous boundary condition of the first or the second kind according as the type of the boundary condition defined in (1.2) on the sides of the polygon Ω on which the indicated parts of the boundary of the blocks lie.

3.2.   Preliminary Lemmas

Let $\gamma_j' = \{(r_j, \theta_j) : r_j = r_{j0}, 0 \le \theta_j \le \alpha_j \pi\}$ be an arc lying on the boundary of the extended block-sector T_j, $1 \le j \le N$, $\gamma_{jp} = (\gamma_{j-p} \cap \overline{T}_j) \setminus (\gamma_j' \cup P_j)$, $p = 0, 1$.

We shall consider an auxiliary boundary-value problem on a sector:

$$\begin{aligned}
\Delta w_j &= 0 && \text{on} \quad T_j, \\
w_j &= f_j(\theta_j) && \text{on} \quad \gamma_j', \\
\nu_{j-p} w_j + \bar{\nu}_{j-p}(w_j)_n' &= 0 && \text{on} \quad \gamma_{jp}, \quad p = 0, 1,
\end{aligned} \tag{3.19}$$

where $\nu_{j-p} = 0$ or 1, $\bar{\nu}_{j-p} = 1 - \nu_{j-p}$ are the same parameters as in the boundary conditions (1.2), $f_j(\theta_j)$ is a given function continuous on γ_j' and, if $\nu_j = 1$, then $f_j(0) = 0$ and if $\nu_{j-1} = 1$, then $f_j(\alpha_j\pi) = 0$.

Lemma 3.1. *There exists a unique solution w_j of the boundary-value problem* (3.19) *satisfying the following requirements:*

(1) *the function w_j is bounded and harmonic on T_j,*

(2) *w_j is continuous up to the boundary of the sector T_j, except, maybe, its vertex P_j,*

(3) *on the sector T_j the function w_j has partial derivatives $\partial w_j/\partial x$ and $\partial w_j/\partial y$, continuous up to its sides γ_{j0} and γ_{j1} (without the end-points),*

(4) *the function w_j satisfies the given boundary conditions on γ_{j0}, γ_{j1} and γ_j'.*

This solution possesses the following additional properties:

(a) *the solution w_j can be represented on $T_j \cup \gamma_{j0} \cup \gamma_{j1}$ in the form*

$$w_j(r_j, \theta_j) = \int_0^{\alpha_j\pi} f_j(\eta) R_j(r_j, \theta_j, \eta)\, d\eta, \qquad (3.20)$$

where R_j is kernel (3.17),

(b) *the definition of the function w_j at the vertex P_j can be extended by continuity (in particular, by formula* (3.20)),

(c) *for $0 < r_j \leq r_{j0} - \delta$, $0 \leq \theta_j \leq \alpha_j\pi$ the inequality*

$$\left|\frac{\partial w_j}{\partial x}\right| + \left|\frac{\partial w_j}{\partial y}\right| \leq c_{j\delta} r_j^{\lambda_j - 1} \qquad (3.21)$$

holds true, where $\delta \in (0, r_{j0}/2]$ is any number, λ_j is quantity (3.18), *$c_{j\delta}$ is a constant independent of r_j and of θ_j.*

Proof.   Let us assume that $\nu_{j-1} = \nu_j$. We consider a function

$$v_j(r, \theta) = \int_0^{\alpha_j\pi} f_j(\eta) R_j(r^{\alpha_j}, \alpha_j\theta, \eta)\, d\eta, \qquad (3.22)$$

where R_j is kernel (3.17), $0 \le r < r_{j0}^{1/\alpha_j}$, $-\pi < \theta \le \pi$. Taking (3.12), (3.17), (3.18) into account, we transform expression (3.22) as follows:

$$v_j(r,\theta) = \frac{1}{\alpha_j} \int\limits_0^{\alpha_j\pi} f_j(\eta) R\left(\frac{r}{r_{j0}^{1/\alpha_j}}, \theta, \frac{\eta}{\alpha_j}\right) d\eta$$

$$+ (-1)^{\nu_j} \frac{1}{\alpha_j} \int\limits_0^{\alpha_j\pi} f_j(\eta) R\left(\frac{r}{r_{j0}^{1/\alpha_j}}, \theta, \frac{-\eta}{\alpha_j}\right) d\eta$$

$$= \int\limits_{-\pi}^{\pi} F_j(\xi) R\left(\frac{r}{r_{j0}^{1/\alpha_j}}, \theta, \xi\right) d\xi, \tag{3.23}$$

where R is kernel (3.14),

$$F_j(\xi) = \begin{cases} f_j(\alpha_j \xi), & 0 \le \xi \le \pi, \\ (-1)^{\nu_j} f_j(-\alpha_j \xi), & -\pi \le \xi < 0, \end{cases}$$

i.e., function (3.22) can be represented as the well-known Poisson integral for a circle.

Since the function $F_j(\xi)$ is continuous on the interval $[-\pi, \pi]$ and $F(-\pi) = F(\pi)$, the function $v_j(r,\theta)$ is harmonic on the open disk

$$T = \{(r,\theta) : 0 \le r < r_{j0}^{1/\alpha_j}, \ -\pi < \theta \le \pi\}$$

and continuous up to the boundary of the disk T on which it assumes values coincident with $F_j(\theta)$ (see [23]). Since the function $F_j(\xi)$ is odd for $\nu_j = 1$ (even for $\nu_j = 0$) with respect to the point $\xi = 0$, the function $v_j(r,\theta)$ is odd (even) with respect to the diameter D connecting the points $(r_{j0}^{1/\alpha_j}, 0)$ and $(r_{j0}^{1/\alpha_j}, \pi)$. Consequently, the function $v_j(r,\theta)$ (the derivative of the function v_j in the direction orthogonal to the diameter D) is zero at the intersection of this diameter D and the disk T for $\nu_j = 1$ (for $\nu_j = 0$).

It easily follows from these properties of the function $v_j(r,\theta)$ that for $\nu_{j-1} = \nu_j$ the function

$$w_j(r_j, \theta_j) = v_j\left(r_j^{1/\alpha_j}, \frac{\theta_j}{\alpha_j}\right) \tag{3.24}$$

is a solution of the boundary-value problem (3.19) satisfying requirements (1)–(4). In addition, function (3.24) possesses properties (a),
(b), (c). Indeed, property (a) follows from (3.22) and (3.24). Property (b) is obvious. And, finally, property (c), i.e., inequality (3.21),
can be established by differentiation of expression (3.20) under the
integral sign for $0 < r_j \le r_{j0} - \delta$ with due account of (3.12), (3.14),
(3.17).

For $\nu_{j-1} \ne \nu_j$ we consider a function

$$v_j(r, \theta) = \int_0^{\alpha_j \pi} f_j(\eta) R_j(r^{2\alpha_j}, 2\alpha_j \theta, \eta) \, d\eta, \qquad (3.25)$$

where R_j is kernel (3.17), $0 \le r < r_{j0}^{1/2\alpha_j}$, $-\pi < \theta \le \pi$. By analogy
with (3.23), bearing in mind (3.12)–(3.14), (3.17), (3.18), we can
reduce expression (3.25) to the form

$$v_j(r, \theta) = \int_{-\pi}^{\pi} F_j(\xi) R\left(\frac{r}{r_{j0}^{1/2\alpha_j}}, \theta, \xi\right) d\xi, \qquad (3.26)$$

where R is kernel (3.14) of the Poisson integral for a circle,

$$F_j(\xi) = \begin{cases} f_j(2\alpha_j \xi), & 0 \le \xi \le \pi/2, \\ (-1)^{\nu_j} f_j(-2\alpha_j \xi), & -\pi/2 \le \xi < 0, \\ (-1)^{\nu_j - 1} f_j(2\alpha_j(\pi - \xi)), & \pi/2 < \xi \le \pi, \\ -f_j(2\alpha_j(\pi + \xi)), & -\pi \le \xi < -\pi/2. \end{cases} \qquad (3.27)$$

In this case the function $F_j(\xi)$ is continuous on the interval $[-\pi, \pi]$
and $F(-\pi) = F(\pi)$. Consequently, function (3.25) is harmonic on
the open disk

$$T = \{(r, \theta) : 0 \le r < r_{j0}^{1/2\alpha_j}, -\pi < \theta \le \pi\}$$

and continuous up to the boundary of the disk T on which it assumes
values coincident with $F_j(\theta)$.

Let us denote by $D(\beta)$ the diameter of the disk T which makes
an angle β with the ray $\theta = 0$. Since function (3.27), extended with
period 2π to the entire ξ-axis, is odd for $\nu_j = 1$ (even for $\nu_j = 0$)
with respect to the point $\xi = 0$ and even (odd) with respect to the

point $\xi = \pi/2$, it follows, respectively, that function (3.26) is odd for $\nu_j = 1$ (even for $\nu_j = 0$) with respect to the diameter $D(0)$ and even (odd) with respect to the diameter $D(\pi/2)$. By virtue of (3.25) and what was said above, the function

$$w_j(r_j, \theta_j) = v_j\left(r_j^{1/2\alpha_j}, \frac{\theta_j}{2\alpha_j}\right) \tag{3.28}$$

is a solution of the boundary-value problem (3.19) for $\nu_{j-1} \ne \nu_j$ which satisfies requirements (1)–(4) and possesses properties (a) and (b).

For $\nu_{j-1} \ne \nu_j$, property (c) can be established by analogy with the preceding case $\nu_{j-1} = \nu_j$. We have thus proved the existence of the required solution of the boundary-value problem (3.19).

To prove that this solution is unique, we consider the corresponding homogeneous boundary-value problem, i.e., set $f_j(\theta_j) \equiv 0$. We assume that the homogeneous problem has a solution $w_j^0 = w_j^0(r_j, \theta_j)$, which satisfies requirements (1)–(4), and

$$\sup_{(r_j, \theta_j) \in T_j \cup \gamma_{j0} \cup \gamma_{j1}} |w_j^0| > 0.$$

Let us assume that $\nu_{j-1} = \nu_j$ and consider a function

$$v_j(r, \theta) = w_j^0(r^{\alpha_j}, \alpha_j \theta), \tag{3.29}$$

which is evidently bounded and harmonic on the open half-disk

$$T' = \{(r, \theta) : 0 < r < r_{j0}^{1/\alpha_j}, 0 < \theta < \pi\},$$

and continuous on the closed half-disk $\overline{T}'$, except, maybe, for the midpoint of the diameter D which supports the half-disk T'. For $\nu_j = 1$ function (3.29) vanishes at all points of the diameter D, except, maybe, its midpoint P_j'. For $\nu_j = 0$ there exists and vanishes a derivative of the function v_j along the inner normal to D at all points of the diameter D, except, maybe, its endpoints and the point P_j'. Consequently, if we oddly extend the function v_j for $\nu_j = 1$ (evenly for $\nu_j = 0$) from T' across the diameter D to a disk with a punctured center, i.e.,

$$T = \{(r, \theta) : 0 < r < r_{j0}^{1/\alpha_j}, -\pi < \theta \le \pi\},$$

then the extended function, for which we retain the designation v_j, will be zero on the boundary of the disk, harmonic and bounded on T. Thus the function v_j has a removable singularity at the center of the disk. Therefore $v_j \equiv 0$ on T, i.e., in accordance with (3.29),

$$w_j^0(r_j, \theta_j) \equiv v_j(r_j^{1/\alpha_j}, \theta_j/\alpha_j) \equiv 0,$$

$$0 < r_j \leq r_{j0}, \qquad 0 \leq \theta_j \leq \alpha_j \pi.$$

Hence the solution of the nonhomogeneous boundary-value problem (3.19), satisfying requirements (1)–(4), is unique for $v_{j-1} = v_j$.

For $v_{j-1} \neq v_j$ the uniqueness can be established by analogy. We have proved Lemma 3.1.

Remark 3.5. Since the solution of the boundary-value problem (3.19), satisfying requirements (1)–(4) in Lemma 3.1, is unique and necessarily possesses property (b), we understand the solution of problem (3.19) to be a function continuous up to the whole boundary of the sector T_j (including the vertex P_j).

Remark 3.6. If $v_{j-1} + v_j \leq 1$, $\alpha_j = 2$ and the values of the function $f_j(\theta_j)$ do not coincide for $\theta_j = 0$ and $\theta_j = \alpha_j \pi$, i.e., on the overlapping ends of the arc γ_j', then the limiting values of the function $w_j(r_j, \theta_j)$ do not coincide either as $r_j \to r_{j0} - 0$, $\theta_j \to +0$ and $r_j \to r_{j0} - 0$, $\theta_j \to 2\pi - 0$. For $\alpha_j = 2$ the sector T_j is a disk with a cut passing from the center in the direction of the ray $\theta_j = 0$ ($\theta_j = 2\pi$). The sides γ_{jp}, $p = 0, 1$, of the sector T_j lie precisely on this cut.

Suppose that r, θ is a polar coordinate system chosen arbitrarily in the complex plane z,

$$T' = \{(r, \theta) : 0 < r < r_0, \ 0 < \theta < \pi\}$$

is a half-disk,

$$\gamma' = \{(r, \theta) : r = r_0, \ 0 \leq \theta \leq \pi\}$$

is an arc (the curvilinear part of the boundary of the half-disk T'), $D' = \overline{T'} \setminus (T' \cup \gamma')$ is the diameter supporting the half-disk T'.

Let us consider a boundary-value problem on a half-disk:

$$\begin{aligned} \Delta w &= 0 &&\text{on}\quad T', \\ w &= f(\theta) &&\text{on}\quad \gamma', \\ \nu w + \overline{\nu} w_n' &= 0 &&\text{on}\quad D', \end{aligned} \qquad (3.30)$$

where $\nu = 0$ or 1, $\bar{\nu} = 1 - \nu$, w'_n is the derivative along the inner normal to D', $f(\theta)$ is a function defined and continuous on γ', and if $\nu = 1$, then $f(0) = f(\pi) = 0$.

Lemma 3.2. *There exists a unique solution w of the boundary-value problem (3.30) which is continuous on $\overline{T}'$ and has on T' partial derivatives $\partial w/\partial x$ and $\partial w/\partial y$ continuous up to D', and this solution can be represented on $T' \cup D'$ as*

$$w(r,\theta) = \int\limits_0^\pi f(\eta) R\left(\nu, \nu, \frac{r}{r_0}, \theta, \eta\right) d\eta,$$

where R is kernel (3.12).

Lemma 3.2 can be proved by analogy with Lemma 3.1.

3.3. Representation of the Solution on Blocks

The following theorem plays an important part in the construction and substantiation of the block method.

Theorem 3.1. *The solution u of the boundary-value problem (1.1), (1.2) can be represented on $\overline{T}_\tau \setminus \gamma'_\tau$, $1 \le \tau \le M$, as*

$$u(r_\tau, \theta_\tau) = Q_\tau(r_\tau, \theta_\tau)$$
$$+ \int\limits_0^{\alpha_\tau \pi} (u(r_{\tau 0}, \eta) - Q_\tau(r_{\tau 0}, \eta)) R_\tau(r_\tau, \theta_\tau, \eta) \, d\eta, \quad (3.31)$$

where T_τ is an extended block (a sector of a disk, a half-disk or a disk), γ'_τ is the curvilinear part of its boundary, $u(r_{\tau 0}, \theta_\tau)$ is the trace of the solution u on γ'_τ, Q_τ and R_τ is a carrier function and a kernel defined by formulas (3.2), (3.4), (3.6), (3.9)–(3.11), (3.15)–(3.17), $\alpha_j \pi$, $1 \le j \le N$, is an interior angle of the polygon Ω with vertex P_j, $\alpha_q = 1$, $N < q \le L$, $\alpha_p = 2$, $L < p \le M$.

Proof.   For $L < \tau \le M$, by virtue of (3.11), (3.14), (3.15) and the relation $\alpha_\tau = 2$, expression (3.31) is the Poisson integral for the disk T_τ. Since the solution u is harmonic on the extended disk and is continuous on $\overline{T}_\tau$ according to Theorem 1.1 and to requirements I–III (Sec. 2) imposed on the covering of the polygon Ω, relation (3.31) is valid for $L < \tau \le M$.

We set

$$u_\tau(r_\tau,\theta_\tau) = u(r_\tau,\theta_\tau) - Q_\tau(r_\tau,\theta_\tau) \qquad (3.32)$$

on $\overline{T}_\tau$. If T_τ is a sector with a vertex angle equal to 2π, and, consequently, the sides $\gamma_{\tau-1}$ and γ_τ of the polygon Ω are superimposed (lie on the cut), then we understand the values of the function u_τ on $\gamma_{\tau-p} \cap \overline{T}_\tau$, $p = 0, 1$, to be limiting values from T_τ as viewed from the inner normal to $\gamma_{\tau-p}$.

Let $1 \le \tau \le N$. Then, according to Theorem 1.1 and Remark 3.1, function (3.32) is harmonic and bounded on the open extended sector T_τ, continuous and continuously differentiable up to the boundary of the sector T_τ, except, maybe, for its vertex P_τ. On $\gamma_{\tau-p} \cap \overline{T}_\tau \setminus P_\tau$, $p = 0, 1$, for $\nu_{\tau-p} = 1$, (for $\nu_{\tau-p} = 0$) this function satisfies the homogeneous boundary condition of the first (second) kind, and if $\nu_\tau = 1$, then $u(r_{\tau 0}, 0) - Q_\tau(r_{\tau 0}, 0) = 0$ and, moreover, if $\nu_{\tau-1} = 1$, then $u(r_{\tau 0}, \alpha_\tau \pi) - Q_\tau(r_{\tau 0}, \alpha_\tau \pi) = 0$. Consequently, according to Lemma 3.1 and Remarks 3.2, 3.5, the function $u_\tau(r_\tau, \theta_\tau)$ is continuous up to the vertex P_τ and can be represented as

$$u_\tau(r_\tau,\theta_\tau) = \int\limits_0^{\alpha_\tau\pi} (u(r_{\tau 0},\eta) - Q_\tau(r_{\tau 0},\eta)) R_\tau(r_\tau,\theta_\tau,\eta)\, d\eta, \qquad (3.33)$$

where $0 \le r_\tau < r_{\tau 0}$, $0 \le \theta_\tau \le \alpha_\tau \pi$. This relation and (3.32) yield representation (3.31) for $1 \le \tau \le N$.

For $N < \tau \le L$ representation (3.31) can be established by analogy, on the basis of Theorem 1.1 and Lemma 3.2 with due account of Remark 3.3. We have proved Theorem 3.1.

Remark 3.7. It is natural to call the number $\alpha_\tau \pi$ the angular magnitude of the block T_τ (T_τ^*). Indeed, for $1 \le \tau \le N$, $\alpha_\tau \pi$ is the magnitude of the vertex angle of the sector T_τ, for $N < \tau \le L$ $\alpha_\tau = 1$, i.e., $\alpha_\tau \pi = \pi$ and the block T_τ is a half-disk, and for $L < \tau \le M$ $\alpha_\tau = 2$, i.e., $\alpha_\tau \pi = 2\pi$ and the block T_τ is a disk.

Remark 3.8. If P_τ is the vertex of an angle of the polygon Ω with $\nu_{\tau-1} = \nu_\tau = 1$, and the boundary values defined in (1.2) on the sides $\gamma_{\tau-1}$ and γ_τ are noncoincident at P_τ, then both the solution u of problem (1.1), (1.2) and the carrier function Q_τ, defined by formula (3.2) for $j = \tau$, are discontinuous at this vertex. However, their difference $u - Q_r$, i.e., function (3.32), is continuous at the vertex P_r according to Lemma 3.1 and, evidently, vanishes at it. This implies that the solution u and the function Q_r come infinitely close to each other near the vertex P_r, although they have no limit at P_r. Thus the carrier function Q_r, which defines the term before the integral sign in representation (3.31), absorbs the singularity, arising in the solution u in connection with the discontinuity of the boundary values at the vertex P_r. The second term in (3.31), i.e., the integral, is a continuous function on $(\overline{T}_r \setminus \gamma'_r) \supset \overline{T}^*_r \ni P_r$.

3.4.   Additional Lemmas

We shall prove the lemma which we shall need later.

Lemma 3.3. *If the specified boundary values are continuous at all vertices of the angles of the polygon Ω on both sides of which the boundary conditions (1.2) are conditions of the first kind, then the equality*

$$\sum_{j=1}^{N} \int_{\gamma_j} u'_n \, ds = 0 \qquad (3.34)$$

is valid for the solution u of problem (1.1), (1.2). Here u'_n is the derivative of the solution u along the inner normal to the side γ_j, s is the arc length increasing in the direction of the positive traverse of the boundary. The integrals under the sign of the sum are, in general, improper convergent integrals with singularities at the endpoints of the sides.

Proof.   We set (see (2.1))

$$T_{j\varepsilon} = T_j(\varepsilon r_{j0}), \qquad 0 < \varepsilon \leq 1, \qquad 1 \leq j \leq N.$$

We have $T_{j1} = T_j$, $1 \leq j \leq N$, i.e., for $\varepsilon = 1$ the sector $T_{j\varepsilon}$ coincides with the extended sector T_j belonging to the specified covering of the polygon Ω (see Sec. 2). For $0 < \varepsilon < 1$ the sector $T_{j\varepsilon}$ constitutes a part of the sector T_j. Evidently, there is a sufficiently small $\varepsilon_0 \in (0, 1/2]$ such that the condition

$$\overline{T}_{j\varepsilon} \cap \overline{T}_{k\varepsilon} \cap \Omega = \emptyset, \qquad j \neq k,$$

is satisfied for any $\varepsilon \in (0, \varepsilon_0]$. In this case the set

$$\Omega_\varepsilon = \Omega \setminus \bigcup_{j=1}^{N} \overline{T}_{j\varepsilon}$$

is connected, i.e., Ω_ε is a domain. In what follws we shall restrict our discussion to the values $\varepsilon \in (0, \varepsilon_0)$.

Suppose that $\sigma_{j\varepsilon}$ is the curvilinear part of the boundary of the sector $T_{j\varepsilon}$, $\gamma_{j\varepsilon}$ is the set of points of the side γ_j attainable from the side of the inner normal to γ_j. Evidently, the boundary of the domain Ω_ε consists of segments (straight lines) $\gamma_{j\varepsilon}$ and arcs (circles) $\sigma_{j\varepsilon}$, $1 \le j \le N$, i.e., it is piecewise-smooth. Consequently, since in accordance with Theorem 1.1 (see also Definition 1.5) the solution u of the boundary-value problem (1.1), (1.2) is harmonic on Ω_ε and continuously differentiable up to the boundary of the domain Ω_ε, it follows, by virtue of Green's formula, that the equality

$$\sum_{j=1}^{N} \int_{\gamma_{j\varepsilon}} u'_n \, ds + \sum_{j=1}^{N} \int_{\sigma_{j\varepsilon}} u'_n \, d\sigma = 0 \tag{3.35}$$

is valid, where u'_n is the derivative of the solution u along the normal which on $\gamma_{j\varepsilon}$ coincides with the inner normal to γ_j and on $\sigma_{j\varepsilon}$ is directed toward Ω_ε, $d\sigma$ is an element of the arc length reckoned along $\sigma_{j\varepsilon}$.

On the basis of Theorem 3.1, the solution u can be represented on the semi-open sector

$$T'_j = \{(r_j, \theta_j) : 0 \le r_j < r_{j0}, \, 0 \le \theta_j \le \alpha_j \pi\}$$

in form (3.31), where $\tau = j$, $1 \le j \le N$. On the assumptions we have made, according to Theorem 1.1, the solution u and the carrier function Q_j are continuous at the vertex P_j, $1 \le j \le N$. Therefore, for $\nu_{j-1} = \nu_j = 1$, expression (3.2) does not contain the second term, and, consequently, the (rough) estimate

$$\left| \frac{\partial Q_j}{\partial x} \right| + \left| \frac{\partial Q_j}{\partial y} \right| = O(r_j^{-3/4}) \tag{3.36}$$

is valid for the first derivatives of the carrier function Q_j which can be found from any one of formulas (3.2), (3.4), (3.6), (3.9). By virtue of (3.32), (3.36) and Lemma 3.1 (inequality (3.21)), with due account of the inequalities $\lambda_j \geq 1/2\alpha_j \geq 1/4$, and Theorem 1.1, which guarantees the continuity of u'_n on every side γ_j, except, maybe, its endpoints, we find that

$$\sup_{s \in \gamma_{je}} |u'_n| + \sup_{\sigma \in \sigma_{je}} |u'_n| = O(\varepsilon^{-3/4}), \qquad 1 \leq j \leq N.$$

Bearing in mind that the length of the curve σ_{je} is $\alpha_j \pi r_{j0}\varepsilon$, we infer that the second sum on the left-hand side of equality (3.35) tends to zero as $\varepsilon \to +0$ and the integrals entering into the first sum tend to the values which, in general, correspond to the improper convergent integrals appearing under the sign of the sum in (3.34). Thus, in the limit, as $\varepsilon \to +0$, relation (3.35) passes into equality (3.34). We have proved Lemma 3.3.

Remark 3.9. For $\bar{\nu}_j = 1$ the integral $\int_{\gamma_j} u'_n \, ds$ under the sign of the sum in equality (3.34) can be replaced by the integral $\int_{\gamma_j} \varphi_j(s) \, ds$ by virtue of item (3) of Definition 1.5 and the boundary conditions (1.2).

We shall formulate one more useful lemma.

Lemma 3.4. *Suppose that the function w is harmonic in the open disk K_ρ of radius ρ lying in the complex plane z, $z = x + iy$, and continuous on the closed disk $\overline{K}_\rho$, with $|w| \leq A$ on $\overline{K}_\rho$. Then, the inequality*

$$\left| \frac{\partial^p w}{\partial x^{p-q} \partial y^q} \right| \leq \frac{p! 2^{p+2} A}{\pi \rho^p},$$

where $p = 1, 2, \ldots$, $0 \leq q \leq p$, is satisfied at the center of the disk K_ρ.

Lemma 3.4 follows from estimates (29), (30) of paper [48] and also from Lemma 9.1 in [67].

4. An Algebraic Problem

We assume that the boundary-value problem (1.1), (1.2) is posed on a finitely-connected polygon Ω. A covering by a finite number of blocks (sectors of disks, half-disks, disks) is constructed for this polygon in accordance with requirements I–IV formulated in Sec. 2. A carrier function Q_μ is defined by one of formulas (3.2), (3.4), (3.6), (3.9)–(3.11) on every closed extended block $\overline{T}_\mu$, $1 \leq \mu \leq M$. In addition, the kernels R_μ, $1 \leq \mu \leq M$, are defined by formulas (3.12)–(3.18).

We introduce a natural number parameter n (the principal parameter of the method) and the quantities

$$n_\mu = \max\{4, [\alpha_\mu n]\}, \tag{4.1}$$

$$\beta_\mu = \alpha_\mu \pi / n_\mu, \tag{4.2}$$

$$\theta_\mu^k = (k - 1/2)\beta_\mu, \tag{4.3}$$

where $1 \leq \mu \leq M$, $1 \leq k \leq n_\mu$, $[\,]$ is the sign of the integer part and, as before, $\alpha_p = 2$ for $L < p \leq M$, $\alpha_q = 1$ for $N < q \leq L$ and $\alpha_j \pi$, $0 < \alpha_j \leq 2$, $1 \leq j \leq N$, is the magnitude of the interior angle made by the sides γ_{j-1} and γ_j of the polygon Ω.

Let us consider a point P_μ^k which lies on the curvilinear part of the boundary of the extended block T_μ and has polar coordinates $r_\mu = r_{\mu 0}$, $\theta_\mu = \theta_\mu^k$, $0 < \theta_\mu^k < \alpha_\mu \pi$. By virtue of condition IV imposed on the covering of the polygon Ω (see Sec. 2), this point falls on some closed basic block $\overline{T}_\tau^*$ with τ known to be unequal to μ, $(\tau \neq \mu)$ (see Fig. 25). For definiteness, let

$$\tau = \tau(\mu, k) = \min\{q : P_\mu^k \in \overline{T}_q^*, 1 \leq q \leq M\}. \tag{4.4}$$

Denoting by $r_{\tau\mu}^k$, $\theta_{\tau\mu}^k$ the coordinates of the point P_μ^k in the polar coordinate system r_τ, θ_τ (see Fig. 25), we can write an equality

$$u(r_{\tau\mu}^k, \theta_{\tau\mu}^k) = Q_\tau(r_{\tau\mu}^k, \theta_{\tau\mu}^k) + \int_0^{\alpha_\tau \pi} (u(r_{\tau 0}, \eta) - Q_\tau(r_{\tau 0}, \eta)) R_\tau(r_{\tau\mu}^k, \theta_{\tau\mu}^k, \eta) \, d\eta,$$

$$1 \leq \mu \leq M, \qquad 1 \leq k \leq n_\mu, \tag{4.5}$$

on the basis of Theorem 3.1. On the left-hand side we have the solution u of problem (1.1), (1.2) at the point P_μ^k which simultaneously

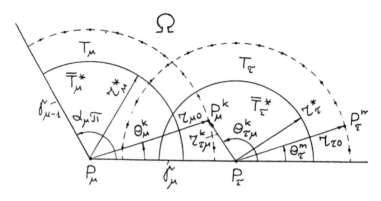

Figure 25

lies on the curvilinear part of the boundary of the extended block T_μ and on the closed basic block $\overline{T}_\tau^*$. On the right-hand side, under the integral sign, there are the values of the solution u on the curvilinear part of the boundary of the extended block T_τ for which $\tau = \tau(\mu, k)$ is defined by formula (4.4). Thus we have written equality (4.5) for each of the n_μ points P_μ^k $(1 \le k \le n_\mu)$ arranged with a step β_μ on the curvilinear part of the boundary of each of the M extended blocks T_μ $(1 \le \mu \le M)$.

Points P_τ^m, which have polar coordinates $r_\tau = r_{\tau 0}$, $\theta_\tau = \theta_\tau^m = (m - 1/2)\beta_\tau$, $1 \le m \le n_\tau$, are arranged, with the step β_τ, on the arc of the circle along which the integral on the right-hand side of (4.5) is calculated. Approximating the integral in (4.5) by the quadrature formula of rectangles with nodes P_τ^m, $1 \le m \le n_\tau$, we get approximate relations

$$u(r_{\tau\mu}^k, \theta_{\tau\mu}^k) \approx Q_\tau(r_{\tau\mu}^k, \theta_{\tau\mu}^k)$$

$$+\beta_\tau \sum_{m=1}^{n_\tau} (u(r_{\tau 0}, \theta_\tau^m) - Q_\tau(r_{\tau 0}, \theta_\tau^m)) R_\tau(r_{\tau\mu}^k, \theta_{\tau\mu}^k, \theta_\tau^m),$$

$1 \le \mu \le M, 1 \le k \le n_\mu, \tau = \tau(\mu, k)$.
     We divide the sum appearing on the right-hand side of these approximate relations by the quantity

$$\max\{1, \beta_\tau \sum_{q=1}^{n_\tau} R_\tau(r^k_{\tau\mu}, \theta^k_{\tau\mu}, \theta^q_\tau)\},$$

close to 1, respectively, (see estimates (6.8), (6.9) below) and introduce notations

$$Q^k_{\tau\mu} = Q_\tau(r^k_{\tau\mu}, \theta^k_{\tau\mu}), \qquad Q^m_\tau = Q_\tau(r_{\tau 0}, \theta^m_\tau), \qquad (4.6)$$

$$R^{km}_{\tau\mu} = \frac{R_\tau(r^k_{\tau\mu}, \theta^k_{\tau\mu}, \theta^m_\tau)}{\max\{1, \beta_\tau \sum_{q=1}^{n_\tau} R_\tau(r^k_{\tau\mu}, \theta^k_{\tau\mu}, \theta^q_\tau)\}}. \qquad (4.7)$$

Then, formally replacing the values $u(r^k_{\tau\mu}, \theta^k_{\tau\mu})$, $u(r_{\tau 0}, \theta^m_\tau)$ of the required solution u at the points P^k_μ, P^m_τ by u^k_μ, u^m_τ and equating the left-hand and right-hand sides of the resulting approximate relations, we arrive at a system of $O(n)$ linear algebraic equations

$$u^k_\mu = Q^k_{\tau\mu} + \beta_\tau \sum_{m=1}^{n_\tau} (u^m_\tau - Q^m_\tau) R^{km}_{\tau\mu}, \qquad (4.8)$$

$$1 \leq \mu \leq M, \qquad 1 \leq k \leq n_\mu, \qquad \tau = \tau(\mu, k),$$

in the unknowns u^k_μ. Evidently, the number of unknowns coincides with the number of equations in the system.

The solution u^k_μ, $1 \leq \mu \leq M$, $1 \leq k \leq n_\mu$, of system (4.8) (provided that it exists) can be accepted as an approximate solution of problem (1.1), (1.2) on the set of points $\{P^k_\mu\}^{k=1,2,\dots,n_\mu}_{\mu=1,2,\dots,M}$ lying on a finite number of arcs of circles (on the curvilinear parts of the boundaries of the extended blocks T_μ) in closed polygon $\overline{\Omega}$. With an increase in n the number of points P^k_μ increases in an approximate proportion to n only on the indicated fixed arcs. Therefore this approximate solution is of no interest. However, with the use of the solution of the algebraic system (4.8) we can construct an approximate solution of problem (1.1), (1.2) in the form of elementary harmonic functions on all basic blocks which are exponentially convergent. We shall discuss this in the next section, and here we shall consider some problems connected with a practical solution of system (4.8) itself.

Lemma 4.1. *There exists a natural n_0 such that for all $n \geq n_0$ the system of linear algebraic equations (4.8) has a unique solution u^k_μ, $1 \leq \mu \leq M$, $1 \leq k \leq n_\mu$.*

Lemma 4.1 is insufficient for practical purposes since n_0 is unknown and, in general, system (4.8) can possess a unique solution for some $n < n_0$. Therefore the criterion of a unique solvability of system (4.8) formulated as Lemma 4.2 below is useful.

Suppose that

$$\varepsilon_{0\mu}^k = 1, \tag{4.9}$$

$$\varepsilon_{\nu\mu}^k = \beta_\tau \sum_{m=1}^{n_\tau} \varepsilon_{\nu-1,\tau}^m R_{\tau\mu}^{km}, \qquad \nu = 1, 2, \ldots, \tag{4.10}$$

$1 \le \mu \le M$, $1 \le k \le n_\mu$, $\tau = \tau(\mu, k) \ne \mu$ is defined by (4.4), $R_{\tau\mu}^{km}$ are numerical coefficients (4.7).

We introduce a norm

$$\|a\| = \max_{1 \le \mu \le M} \max_{1 \le k \le n_\mu} |a_\mu^k|. \tag{4.11}$$

In this case

$$\|\varepsilon_\nu - a\| = \max_{1 \le \mu \le M} \max_{1 \le k \le n_\mu} |\varepsilon_{\nu\mu}^k - a_\mu^k|.$$

Lemma 4.2. *For system (4.8) to be uniquely solvable, it is necessary and sufficient that the condition*

$$\|\varepsilon_{M+1}\| < 1 \tag{4.12}$$

be fulfilled.

Lemma 4.3. *The inequality*

$$\sup_n \|\varepsilon_{M+1}\| = \sigma_0 < 1 \tag{4.13}$$

holds true, where the supremum is taken over all natural n for which system (4.8) is uniquely solvable.

We set

$$a_\mu^k = Q_{\tau\mu}^k - \beta_\tau \sum_{m=1}^{n_\tau} Q_\tau^m R_{\tau\mu}^{km}, \tag{4.14}$$

$$u_{\nu\mu}^k = a_\mu^k + \beta_\tau \sum_{m=1}^{n_\tau} u_{\nu-1,\tau}^m R_{\tau\mu}^{km}, \qquad \nu = 1, 2, \ldots, \tag{4.15}$$

where $u_{0\mu}^k = 0$, $1 \le \mu \le M$, $1 \le k \le n_\mu$, $\tau = \tau(\mu, k)$.

Lemma 4.4. *If condition (4.12) is fulfilled, then*

$$\|\varepsilon_{\nu+1}\| \le \|\varepsilon_\nu\|, \tag{4.16}$$

$$\|\varepsilon_\nu\| \le \|\varepsilon_{M+1}\|^{[\nu/(M+1)]} < \sigma_0^{\nu/(M+1)-1}, \tag{4.17}$$

$$\|u_\nu - u\| \le \frac{\|\varepsilon_\nu\|}{1 - \|\varepsilon_\nu\|}\|u_\nu\| \le c_*\|\varepsilon_\nu\|, \tag{4.18}$$

where $\nu = M + 1, M + 2, \ldots$, u is a solution of system (4.8), u_ν is an approximate solution of system (4.8) found after the νth iteration (4.15), c_* is a constant independent of ν and of n.

The proofs of Lemmas 4.1–4.4 are given in Sec. 6.

To verify the uniqueness of solvability of system (4.8) expressed by inequality (4.12), we have to carry out $M + 1$ iterations by formula (4.10). These iterations are quite similar and are even simpler than those carried out by formula (4.15) for finding a solution of system (4.8). Therefore in practice both iterations can be realized on a computer using the same program.

By virtue of Lemmas 4.3 and 4.4, under condition (4.12), the simple iterations (4.15) converge in a uniform metric to a solution of system (4.8) in a geometric progression with a common ratio not exceeding the value $\sigma_0^{1/(M+1)} < 1$ independent of n.

Lemma 4.4 contains an estimate (the left-hand inequality in (4.18)) of the iteration error in terms of an actually calculated value which also decreases in a geometric progression.

We can also use Seidel's method to find a solution of system (4.8):

$$u_{\nu\mu}^k = a_\mu^k + \beta_\tau \sum_{m=1}^{n_\tau} R_{\tau\mu}^{km} \cdot \begin{cases} u_{\nu-1,\tau}^m, & \tau > \mu, \\ u_{\nu\tau}^m, & \tau < \mu, \end{cases} \tag{4.19}$$

where $u_{0\mu}^k = 0$, $1 \le \mu \le M$, $1 \le k \le n_\mu$, $\tau = \tau(\mu, k) \ne \mu$ (see (4.4)), $\nu = 1, 2, \ldots$, converging in practice more rapidly than the method of simple iterations (4.15). Correspondingly, we use Seidel's method to calculate the quantities

$$\varepsilon_{\nu\mu}^k = \beta_\tau \sum_{m=1}^{n_\tau} R_{\tau\mu}^{km} \cdot \begin{cases} \varepsilon_{\nu-1,\tau}^m, & \tau > \mu, \\ \varepsilon_{\nu\tau}^m, & \tau < \mu, \end{cases} \tag{4.20}$$

where $\varepsilon_{0\mu}^k = 1$, $1 \le \mu \le M$, $1 \le k \le n_\mu$, $\tau = \tau(\mu, k)$, $\nu = 1, 2, \ldots$

When using formulas (4.19), (4.20) for a fixed μ, we sort out k, $1 \le k \le n_\mu$, and then increase μ by unity, and so on.

Lemmas 4.2–4.4 are valid for Seidel's method.

Remark 4.1. In practice, to improve the convergence of the method of simple iterations and Seidel's method, it is expedient to number the basic blocks beginning with those which have parts of the boundary in common with the sides of the polygon Ω on which boundary conditions of the first kind are defined. Then the numbering should be continued to the basic blocks which intersect the blocks already numbered, and so on. The numbers of the extended blocks should correspond to those of the basic blocks. The numbering for which the value $\|\varepsilon_{M+1}\|$ is minimal is preferable.

In what follows the numbering of the blocks accepted in Sec. 2 is preserved.

5.   The Main Result

5.1.   Theorem on the Convergence of the Block Method

Definition 5.1. *Let* u_μ^k, $1 \le \mu \le M$, $1 \le k \le n_\mu$, *be a solution of the system of linear algebraic equations* (4.8).

The function[2]

$$u_{\mu n}(r_\mu, \theta_\mu) = Q_\mu(r_\mu, \theta_\mu) + \beta_\mu \sum_{k=1}^{n_\mu} (u_\mu^k - Q_\mu^k) R_\mu(r_\mu, \theta_\mu, \theta_\mu^k), \qquad (5.1)$$

where the kernel R_μ, *the function* Q_μ *and the quantities* n_μ, β_μ, θ_μ^k, Q_μ^k *are defined in Sec. 3 and 4, respectively, is called an approximate solution of the boundary–value problem* (1.1), (1.2) *on the closed basic block (a sector, half-disk or disk)* $\overline{T}_\mu^*$, $1 \le \mu \le M$.

Remark 5.1. The domain of definition of function (5.1) is wider than the closed basic block $\overline{T}_\mu^*$. This function is also defined on the closed block $\overline{T}_\mu^{**}$, where (see (2.1), (2.2))

$$T_\mu^{**} = T_\mu(r_\mu^{**}), \qquad (5.2)$$

$r_\mu^{**} = (r_{\mu 0} + r_\mu^*)/2$, $1 \le \mu \le M$. The block T_μ^{**} is said to be intermediate since $T_\mu^* \subset T_\mu^{**} \subset T_\mu$, where T_μ^*, T_μ are the basic and

---

[2]Here and in what follows, as before, r_μ, θ_μ are polar coordinates on the block T_μ defined in Sec. 2; n is a natural number parameter of the method introduced in Sec. 4.

the extended block. By virtue of Remarks 3.1–3.5, function (5.1) is bounded and harmonic on the open intermediate block $T_\mu^{**} \supset T_\mu^*$ and continuous up to its boundary, except for the point P_μ, if the block T_μ^{**} is a sector, and the boundary values defined in (1.2) for $\nu_{\mu-1} = \nu_\mu = 1$ are discontinuous at its vertex P_μ. In addition, on the rectilinear parts of the boundary of the intermediate block–half–disks and sectors, except, maybe, the vertex of the latter, functions (5.1) satisfy the boundary conditions defined in (1.2). Intermediate blocks are used to prove some assertions in Sec. 6.

We shall formulate the fundamental theorem which characterizes the block method.

Theorem 5.1. *Suppose that u is a solution of the boundary-value problem* (1.1), (1.2), *$u_{\mu n}$ is its approximate solution* (5.1) *on the closed basic block* $\overline{T}_\mu^*$, $1 \le \mu \le M$. *Then the following statements are valid.*

(1) *The inequality*

$$\left| \frac{\partial^p}{\partial x^{p-q} \partial y^q} (u_{\mu n}(r_\mu, \theta_\mu) - u(r_\mu, \theta_\mu)) \right| \le c_p \exp\{-d_0 n\} \qquad (5.3)$$

is satisfied on $\overline{T}_\mu^*$, *first, for $N < \mu \le M$ when $p \ge 0$, second, for $1 \le \mu \le N$, integer λ_μ and any $\nu_{\mu-1}$ and ν_μ when $p \ge \lambda_\mu$, third, for $1 \le \mu \le N$, $\nu_{\mu-1} = \nu_\mu = 0$ and any λ_μ when $p = 0$.*

(2) *The estimate*

$$\left| \frac{\partial^p}{\partial x^{p-q} \partial y^q} (u_{\mu n}(r_\mu, \theta_\mu) - u(r_\mu, \theta_\mu)) \right| \le c_p r_\mu^{\lambda_\mu - p} \exp\{-d_0 n\} \qquad (5.4)$$

is valid on $\overline{T}_\mu^*$ *for $1 \le \mu \le N$ and any λ_μ if $\nu_{\mu-1} + \nu_\mu \ge 1$, $0 \le p < \lambda_\mu$ or $\nu_{\mu-1} = \nu_\mu = 0$, $1 \le p < \lambda_\mu$.*

(3) *The inequality*

$$\left| \frac{\partial^p}{\partial x^{p-q} \partial y^q} (u_{\mu n}(r_\mu, \theta_\mu) - u(r_\mu, \theta_\mu)) \right| \le c_p r_\mu^{\lambda_\mu - p} \exp\{-d_0 n\} \qquad (5.5)$$

holds true on $\overline{T}_\mu^* \setminus P_\mu$ *for $1 \le \mu \le N$, noninteger λ_μ, and any $\nu_{\mu-1}$ and ν_μ when $p > \lambda_\mu$.*

Everywhere $0 \le q \le p$, λ_μ is quantity (3.18), *$\nu_{\mu-1}$ and ν_μ are parameters entering into the boundary conditions* (1.2), *c_p, $p = 0, 1, \ldots,$*

and $d_0 > 0$ are constants independent of $r_\mu = r_\mu(x, y)$, $\theta_\mu = \theta_\mu(x, y)$ and of n, for which the system of linear algebraic equations (4.8) has a unique solution. [3] In (3) P_μ is the vertex of the block-sector T_μ^*.

The proof of Theorem 5.1 is given in Sec. 6.

Remark 5.2. The approximate solution (5.1), being itself an elementary harmonic function, is constructed very naturally. The carrier function $Q_\mu(r_\mu, \theta_\mu)$, $1 \le \mu \le N$, acquires precisely the logarithmic singularities of the derivatives of the required solution u and the possible discontinuity of the solution at the vertex P_μ of the polygon Ω, that appear because the consistency conditions for the boundary conditions at the vertex are not fulfilled. The power singularities of the derivatives, which are essentially due to the presence of an angle (see [50]), are passed on sufficiently exactly to the approximate solution via kernel R_μ. These properties of the approximate solution are a prerequisite for a high rate of its convergence established by Theorem 5.1.

5.2. Discussion of the Theorem

Proceeding from Theorem 5.1, we can make the following conclusions.

I. On every closed basic block $\overline{T}_\mu^*$ the approximate solution (5.1) converges, at least uniformly, according to the exponential law, to the solution of the boundary-value problem (1.1), (1.2). In this case, for $\nu_{\mu-1} + \nu_\mu \ge 1$ ($1 \le \mu \le N$), the error of the approximate solution (5.1) on the block-sectors $\overline{T}_\mu^*$ decreases with the weight $r_\mu^{\lambda\mu}$, where r_μ is the distance from the running point to the vertex P_μ, and λ_μ is quantity (3.18) (estimate (5.4) for $p = 0$).

II. On closed basic block-half-disks and disks $\overline{T}_\mu^*$, $N < \mu \le M$, the derivatives of the approximate solution, which are of an arbitrary order p, converge uniformly, according to the exponential law, to the corresponding derivatives of the desired solution u (estimate (5.3) for $N < \mu \le M$).

III. On the closed sectors $\overline{T}_j^*$ (except, maybe, their vertex P_j), for

---

[3] According to Lemma 4.1 system (4.8) is uniquely solvable, at least for a sufficiently large n.

which the quantity λ_j, sought from formula (3.18), is an integer, the uniform convergence, according to the exponential law, of derivatives of any order is also guaranteed, $1 \le j \le N$. The derivatives of the desired solution u and of the approximate solution u_{jn} may be infinite in the vicinity of the vertex P_j. However, the derivatives of the difference $u_{jn} - u$ not only do not increase in the vicinity of the vertex but even tend to zero at the vertex P_j for $p < \lambda_j$ (estimates (5.3), (5.4), where $\mu = j$).

The number λ_j is natural in the following cases

(a) $\nu_{j-1} = \nu_j$ ($\nu_{j-1}\nu_j + \bar{\nu}_{j-1}\bar{\nu}_j = 1$), i.e., boundary conditions of the same kind are defined on the sides γ_{j-1} and γ_j and α_j assumes one of the values $1, 1/2, 1/3, \ldots$ (the vertex angle of the sector T_j^* has the magnitude π/m, where m is natural, in particular, $m = 1$). The vertex angle P_j equal to π has sense in this case as well, say, when the polynomial entering into the right-hand side of the boundary condition varies jumpwise at the point P_j;

(b) $\nu_{j-1} \ne \nu_j$ ($\nu_{j-1}\nu_j + \bar{\nu}_{j-1}\bar{\nu}_j = 0$), i.e., boundary conditions of different kinds are defined on the sides γ_{j-1} and γ_j and α_j assumes one of the values $1/2, 1/4, 1/6, \ldots$ (the vertex angle of the sector T_j^* is equal to $\pi/2m$, where m is a natural number).

An angle equal to π does not yield an integer value of λ_j when the boundary conditions defined on its sides are of different kinds.

When λ_j is an integer, the derivatives of order p of the solution u of problem (1.1), (1.2) and the approximate solution (5.1) (where $\mu = j$), exist at the vertex P_j of the closed block-sector $\bar{T}_j^*$ and are bounded in its neighborhood only if the following consistency conditions for the boundary conditions are fulfilled (see [50] and the expressions for the carrier functions Q_j in Sec. 3):

(1) for $\nu_{j-1} = \nu_j = 1$ (boundary conditions of the first kind are defined on γ_{j-1} and γ_j) and $\alpha_j = 1/m$ we have

$$a_{j,ml} = (-1)^l b_{j,ml}, \qquad 0 \le l \le \left[\frac{p}{m}\right]; \qquad (5.6)$$

(2) for $\nu_{j-1} = \nu_j = 0$ (boundary conditions of the second kind are defined on γ_{j-1} and γ_j) and $\alpha_j = 1/m$ we have

$$a_{j,ml-1} = (-1)^{l+1} b_{j,ml-1}, \qquad 1 \le l \le \left[\frac{p}{m}\right]; \qquad (5.7)$$

(3) for $\nu_{j-1} = 0$, $\nu_j = 1$ (a boundary condition of the second kind is defined on γ_{j-1} and that of the first kind is defined on γ_j) and $\alpha_j = 1/2m$ we have

$$a_{j,(2l+1)m-1} = (-1)^l(2l+1)mb_{j,(2l+1)m}, \quad 0 \le l \le \left[\frac{p}{2m} - \frac{1}{2}\right]; \quad (5.8)$$

(4) for $\nu_{j-1} = 1$, $\nu_j = 0$ (a boundary condition of the first kind is defined on γ_{j-1} and that of the second kind is defined on γ_j) and $\alpha_j = 1/2m$ we have

$$(2l+1)ma_{j,(2l+1)m} = (-1)^l b_{j,(2l+1)m-1}, \quad 0 \le l \le \left[\frac{p}{2m} - \frac{1}{2}\right], \quad (5.9)$$

where a_{jk}, b_{jk} are coefficients of polynomials (3.1), and, if the index k of a_{jk} or of b_{jk} exceeds the degree of the corresponding polynomial, then this coefficient is naturally zero.

The fulfillment of consistency conditions (5.6)–(5.9) is not required in Theorem 5.1 since it is not necessary. The derivatives of any order of the difference of the solution of boundary-value problem (1.1), (1.2) and of the approximate solution on $\overline{T}_j^*$ for $1 \le j \le N$ and a natural number λ_j have no singularities at the vertex P_j of the closed sector $\overline{T}_j^*$ even if the consistency conditions for the boundary conditions at P_j are not fulfilled.

IV. When λ_j is not an integer, the right-hand sides of the estimates of errors (5.4), (5.5) include a weight factor $r_j^{\lambda_j-p}$, which tends to zero for $p < \lambda_j$ at the vertex P_j and increases indefinitely for $p > \lambda_j$. Thus the error of the derivatives of order $p < \lambda_j$ tends to zero when approaching the vertex P_j even if for $\nu_{j-1} = \nu_j = 1$ the given boundary values are discontinuous at this vertex. For $p > \lambda_j$ the factor $r_j^{\lambda_j-p}$ in estimate (5.5) indirectly reflects the power singularity which, as a rule, the pth derivatives of the desired solution u of problem (1.1), (1.2) have even if the boundary conditions on the sides γ_{j-1} and γ_j are homogeneous (see [50]).

Note that λ_j is greater the smaller the interior angle $\alpha_j\pi$ at the vertex P_j of the polygon Ω (see (3.18)).

Remark 5.3. The closed basic blocks (sectors, half-disks and disks) $\overline{T}_\mu^*$, which form a covering of the closed polygon $\overline{\Omega}$, intersect.

The approximate solution (5.1) of the boundary-value problem (1.1), (1.2) has been obtained on every basic block $\overline{T}_\mu^*$, $1 \le \mu \le M$, as an elementary function which is harmonic on $\overline{T}_\mu^*$ everywhere, except, maybe the point (vertex) P_μ for $1 \le \mu \le N$. Thus there are several approximate solutions at each intersection of the closed basic blocks which, in general, do not coincide, but are close to one another. By virtue of Theorem 5.1 each of these approximate solutions converges exponentially together with its derivatives, at the intersection of the closed basic blocks to the solution of problem (1.1), (1.2) since the vertices P_μ, $1 \le \mu \le N$, at which all the singularities of the solution of problem (1.1), (1.2) are concentrated, do not enter into the intersections.

To illustrate the accuracy of the method being considered, let us consider an example of Dirichlet's problem for Laplace's equation on a regular pentagon Ω with a side equal to unity (see Fig. 21). Assume that the boundary values on every side are either zero or unity (any combinations are admissible). If the boundary values, in their totality are not a constant, then, in the neighborhood of the vertices of the pentagon at which the given boundary values are even continuous, the required solution of Dirichlet's problem has unbounded second derivatives.

Let us verify this.

Suppose, for definiteness, that $\varphi_1(s) = 0$, $\varphi_5(s) = 0$ and boundary values equal to unity are given on at least one of the sides γ_2, γ_3, γ_4 of the pentagon Ω. Then we can represent the solution u of Dirichlet's problem on a closed sector $\overline{T}_1^*$ of radius $r_1^* = 0.57$ whose vertex angle is equal to $3\pi/5$ ($\alpha_1 = 3/5$) in the form (see [50])

$$u(r_1, \theta_1) = b_1 \left(\frac{r_1}{r_{10}}\right)^{5/3} \sin \frac{5\theta_1}{3} + \sum_{k=2}^{\infty} b_k \left(\frac{r_1}{r_{10}}\right)^{5k/3} \sin \frac{5k\theta_1}{3}. \qquad (5.10)$$

Here

$$b_q = \frac{10}{3\pi} \int_0^{3\pi/5} u(r_{10}, \eta) \sin \frac{5q\eta}{3}\, d\eta,$$

$q = 1, 2, \ldots$, $u(r_{10}, \eta)$ is the trace of the (unknown) solution u on the curvilinear part of the boundary of the extended sector $T_1 \subset \Omega$

of radius $r_{10} = 0.89 > r_1^*$. Since the inequalities $0 < u(r_{10}, \eta) < 1$, $0 < \sin \frac{5\eta}{3} \leq 1$ are satisfied for $0 < \eta < 3\pi/5$ and the functions $u(r_{10}, \eta)$ and $\sin \frac{5\eta}{3}$ are continuous in η, we have $b_1 > 0$, $|b_k| < 2$, k=1,2,.... It follows that the series in (5.10) and the series obtained from it by means one-fold and two-fold term-by-term differentiation with respect to x and y (in particular, with respect to r_1) converge uniformly on the closed sector $\overline{T}_1^*$ of radius $r_1^* < r_{10}$. Consequently, the function represented by this series is twice continuously differentiable with respect to x and y on $\overline{T}_1^*$. However, after the two-fold differentiation with respect to r_1 a separate term on the right-hand side of (5.10) has a singularity of form $r_1^{-1/3}$ at the vertex P_1 since b_1 is known to be positive in the case being discussed. In particular, if the bisector of the vertex angle of the sector $\overline{T}_1^*$ lies on the y-axis, then the second derivative of the term in question with respect to y coincides on the bisector with the second derivative relative to r_1 for $\theta_1 = 3\pi/10$ and has the indicated singularity. This singularity is noncritical for this approximate method as distinct, for instance, from the method of uniform nets (see [52]).

The sectors T_j^*, T_j, $j = 1, 2, \ldots, 5$, of radii $r_j^* = 0.57$, $r_{j0} = 0.89$, respectively, and the circles T_6^*, T_6 with centers at the center P_6 of the pentagon (see Fig. 21) with radii $r_6^* = 0.42$, $r_{60} = 0.5 \cot(\pi/5)$ form the covering of the pentagon Ω satisfying requirements I–IV from Sec. 2. For simplicity, the extended sectors T_j, $j = 1, 2, \ldots, 5$, are not shown in Fig. 21.

We have found the upper numerical estimate $\overline{\varepsilon}$ of the maximum error of the approximate solution (5.1), $\mu = 1, 2, \ldots, 6$, of Dirichlet's problem being discussed which assumes the following values:

n	50	100	200
$\overline{\varepsilon}$	$2 \cdot 10^{-2}$	$4 \cdot 10^{-7}$	10^{-16}

The exponential convergence is substantiated.

Remark 5.4. When setting up an algebraic problem in Sec. 4, we defined n_μ points P_μ^k, $1 \leq k \leq n_\mu$, on the curvilinear part of the boundary of the extended block T_μ which are associated with n_μ unknowns and equations in system (4.8). On the closed block $\overline{T}_\mu^*$ the approximate solution (5.1) contains n_μ terms. This n_μ can be found from formula (4.1). Roughly speaking, n_μ is proportional to n with the proportionality factor α_μ if we take into account the angular magnitude $\alpha_\mu \pi$ of the block T_μ^*.

In general, we can set[4]

$$n_\mu = \max\{4, [a_\mu n]\}, \qquad 1 \le \mu \le M, \qquad (5.11)$$

where a_μ are arbitrary fixed positive numbers, say, $a_\mu = 1$ for all μ. Lemmas 4.1–4.4 and Theorem 5.1 remain valid but the constants c_p, $d_0 > 0$ in estimates (5.3)–(5.5) may vary.

A question arises concerning the choice, in some sense, of the optimal values of a_μ, $1 \le \mu \le M$, say, those minimizing c_p for a certain p or maximizing d_0. However, the relationship between the quantities a_μ and the constants c_p and d_0 is very complicated and we cannot trace it sufficiently accurately. Therefore we shall restrict our discussian to qualitative considerations which can be employed in practice when the values of a_μ in (5.11) are being chosen. The quantity a_μ must be greater, larger α_μ, i.e., the angular magnitude of the block (and this is taken into account in (4.1)), the larger $r_{\mu 0}$, i.e., the radial magnitude of the extended block and the closer the vertices of the polygon Ω are to the curvilinear part of the boundary of the extended block T_μ since in their vicinity the smoothness of the required solution is usually worse, and, finally, the closer to unity the ratio $r_\mu^*/r_{\mu 0}$ is since when the ratio $r_\mu^*/r_{\mu 0}$ (to be more precise, $r_\tau^*/r_{\tau 0}$) approaches unity, the approximation of the integrals in (4.5) by quadrature formulas upon the passage to the algebraic system (4.8) and of the integeral in (3.31) in the construction of the approximate solution (5.1) becomes less accurate. In the whole, it is natural to bound the value $\max\limits_{1 \le \mu \le M} a_\mu$ from above, say, by the number 2. This bounding is fulfilled, in particular, for $a_\mu = \alpha_\mu$.

It follows from what was said above that we construct a covering for the polygon Ω according to the laws indicated in Sec. 2, we must not define the extended blocks the curvilinear part of whose boundary

---

[4] Here and in (4.1) we have put 4 under the sign of maximum in order to have a guaranteed number of points P_μ^k on the curvilinear part of the boundary of every extended block T_μ.

passes too close to the vertices of the polygon Ω, and should not take the ratio $r_\mu^*/r_{\mu0}$ of the radial dimensions of the basic and the extended block close to unity.

Below, up to Sec. 14 inclusive, we shall define n_μ by formula (4.1).

6.   Proofs of Theorem 5.1 and of Lemmas 4.1–4.4

6.1.   Preliminary Lemmas

Lemma 6.1. *Suppose $f(\eta)$ is an infinitely differentiable function on the η-axis of a period l such that*

$$\max_{\eta \in [0,l]} |f^{(k)}(\eta)| \le b^k k!, \qquad k = 1, 2, \ldots, \tag{6.1}$$

where $b > 0$ does not depend on k. Then, for any natural ν

$$\left| \int_0^l f(\eta)\, d\eta - \frac{l}{\nu} \sum_{p=1}^{\nu} f\left(\frac{2p-1}{2\nu} l\right) \right| \le c \exp\{-d\nu\}, \tag{6.2}$$

where $c, d > 0$ are constants dependent only on b and l.

Proof.   Let us first establish the inequality

$$\inf_{k \ge 1} \frac{\beta^k k!}{\nu^k} \le (1+\beta) e \exp\left\{-\frac{\nu}{\beta e}\right\}, \tag{6.3}$$

where k and ν are natural, $\beta > 0$. For $\nu \le \beta e$ inequality (6.3) is obvious. Let $\nu > \beta e$, $q = [\nu/\beta e]$. We have

$$\inf_{k \ge 1} \frac{\beta^k k!}{\nu^k} \le \inf_{k \ge 1} \left(\frac{\beta k}{\nu}\right)^k \le \left(\frac{\beta q}{\nu}\right)^q = \left(\frac{\beta[\nu/\beta e]}{\nu}\right)^{[\nu/\beta e]} \le$$

$$\le \left(\frac{1}{e}\right)^{[\nu/\beta e]} \le \exp\left\{1 - \frac{\nu}{\beta e}\right\} = e \exp\left\{-\frac{\nu}{\beta e}\right\}.$$

Inequality (6.3) is proved.

According to Theorem 2 from [24]

$$\left| \int_0^l f(\eta)\, d\eta - \frac{l}{\nu} \sum_{p=1}^{\nu} f\left(\frac{2p-1}{2\nu} l\right) \right|$$

$$\le \frac{\pi^2}{\nu^k} \left(\frac{l}{2\pi}\right)^{k+1} \max_{\eta \in [0,l]} |f^{(k)}(\eta)|, \qquad k = 1, 2 \ldots$$

Together with (6.1) and (6.3) this relation implies (6.2) for the constants $c = le(2\pi + bl)/4$, $d = 2\pi/ble$.

Lemma 6.2. *Let a be a number, $|a| < 1$, $b = 1 - |a|$,*

$$\psi(\eta) = (1 + a\cos\eta)^{-1}. \tag{6.4}$$

Then

$$\max_{|\eta| < \infty} |\psi^{(k)}(\eta)| \leq \frac{k!}{b^{k+1}}, \qquad k = 0, 1, \dots.$$

We can prove Lemma 6.2 by expanding the function $\psi(\eta)$ in the powers of $a\cos\eta$.

Lemma 6.3. *Suppose $R(r, \theta, \eta)$ is kernel (3.14), $0 \leq \rho < 1$. Then*

$$\left| \frac{\partial^k R(r, \theta, \eta)}{\partial \eta^k} \right| \leq a_\rho b_\rho^k k!, \tag{6.5}$$

where $k = 0, 1, \dots$, $0 \leq r \leq \rho$, $|\theta| < \infty$, $|\eta| < \infty$, and a_ρ, b_ρ are constants dependent only on ρ.

Proof. Since

$$R(r, \theta, \eta) = (1 - r^2)\psi(\theta - \eta)/2\pi(1 + r^2),$$

where ψ is function (6.4), in which $a = -2r/(1 + r^2)$ is independent of η, $|a| \leq 2\rho/(1 + \rho^2) < 1$ for $0 \leq r \leq \rho$, it follows from this and from Lemma 6.2 that inequality (6.5) holds for $a_\rho = 1/\pi(1 - \rho)$, $b_\rho = 2/(1 - \rho)^2$.

Lemma 6.4. *Let $R(r, \theta, \eta)$ be kernel (3.14), $0 \leq \rho < 1$. Then*

$$\left| 1 - \frac{2\pi}{\nu} \sum_{p=1}^{\nu} R\left(r, \theta, \frac{(2p-1)\pi}{\nu}\right) \right| \leq c_\rho \exp\{-d_\rho\nu\},$$

where $\nu = 1, 2, \dots$, $0 \leq r \leq \rho$, $|\theta| < \infty$, and c_ρ, $d_\rho > 0$ are constants dependent only on ρ.

Proof. Since (see [23])

$$\int_0^{2\pi} R(r, \theta, \eta)\, d\eta = 1, \qquad 0 \leq r \leq \rho, \qquad |\theta| < \infty,$$

and the function $R(r, \theta, \eta)$ is 2π-periodic and infinitely differentiable with respect to η for $0 \leq r \leq \rho$, the statement of Lemma 6.4 follows from Lemmas 6.3 and 6.1.

Below we assume as before that the defined finite covering of the polygon Ω by blocks of three types (sectors of disks, half-disks, and disks) satisfies requirements I–IV formulated in Sec. 2. Then $n_\mu = \max\{4, [\alpha_\mu n]\}$, $\beta_\mu = \alpha_\mu \pi / n_\mu$, P_μ^k is a point with coordinates $r_\mu = r_{\mu 0}$, $\theta_\mu = \theta_\mu^k = (k - 1/2)\beta_\mu$, $1 \leq \mu \leq M$, $1 \leq k \leq n_\mu$, $\tau = \tau(\mu, k)$ is defined by formula (4.4), $r_{\tau\mu}^k$, $\theta_{\tau\mu}^k$ are the coordinates of the point P_μ^k in the polar coordinate system r_τ, θ_τ (see Fig. 25).

We note for the sequel that the inequalities

$$0 \leq R_{\tau\mu}^{km} \leq R_\tau(r_{\tau\mu}^k, \theta_{\tau\mu}^k, \theta_\tau^m) \qquad (6.6)$$

are valid, where $1 \leq \mu \leq M$, $1 \leq k \leq n_\mu$, $\tau = \tau(\mu, k)$, $1 \leq m \leq n_\tau$, $R_{\tau\mu}^{km}$ is quantity (4.7), R_τ is the kernel defined by relations (3.12)–(3.18). These inequalities follow from (4.7) and the nonnegativity of the kernel $R_\tau(r_\tau, \theta_\tau, \eta)$ on the closed basic block $\overline{T}_\tau^*$ for any $\eta \in (0, \alpha_\tau \pi)$ which can be easily established from (3.12)–(3.18) with due account of the fact that $(r_{\tau\mu}^k, \theta_{\tau\mu}^k) \in \overline{T}_\tau^*$ (see Fig. 25) and $0 < \theta_\tau^m < \alpha_\tau \pi$.

Lemma 6.5. *Let $1 \leq \mu \leq M$, $1 \leq k \leq n_\mu$, $\tau = \tau(\mu, k)$. Then*

$$\beta_\tau \sum_{m=1}^{n_\tau} (R_\tau(r_{\tau\mu}^k, \theta_{\tau\mu}^k, \theta_\tau^k) - R_{\tau\mu}^{km}) \leq c_\tau' \exp\{-d_\tau' n\}, \qquad (6.7)$$

where c_τ', $d_\tau' > 0$ are constants dependent only on the ratio $r_\tau^*/r_{\tau 0}$ and on α_τ.

Proof.   According to (3.12)–(3.18) we have

$$\beta_\tau \sum_{m=1}^{n_\tau} R_\tau(r_{\tau\mu}^k, \theta_{\tau\mu}^k, \theta_\tau^m) \leq \frac{2\pi}{n_\tau^*} \sum_{p=1}^{n_\tau^*} R\left(\left(\frac{r_{\tau\mu}^k}{r_{\tau 0}}\right)^{\lambda_\tau}, \lambda_\tau \theta_{\tau\mu}^k, \frac{(2p - 1)\pi}{n_\tau^*}\right),$$

$$(6.8)$$

where $n_\tau^* = n_\tau$ for $L < \tau \leq M$, $n_\tau^* = 2n_\tau$ for $N < \tau \leq L$ or for $1 \leq \tau \leq N$ and $\nu_{\tau-1} = \nu_\tau$, $n_\tau^* = 4n_\tau$ for $1 \leq \tau \leq N$ and $\nu_{\tau-1} \neq \nu_\tau$; $\lambda_\tau = 1$ for $N < \tau \leq M$, λ_τ is quantity (3.18) for $1 \leq \tau \leq N$. Since $n_\tau = \max\{4, [\alpha_\tau n]\} \geq 4\alpha_\tau n/5$, where $\alpha_\tau = 2$ for $L < \tau \leq M$, $\alpha_\tau = 1$

for $N < \tau \leq L$, $0 < \alpha_\tau \leq 2$ ($\alpha_\tau \pi$ is the magnitude of the angle with vertex P_τ) for $1 \leq \tau \leq N$, and

$$(r^k_{\tau\mu}/r_{\tau 0})^{\lambda_\tau} \leq (r^*_\tau/r_{\tau 0})^{\lambda_\tau} \leq (r^*_\tau/r_{\tau 0})^{1/2\alpha_\tau} < 1$$

by condition (4.4), this result and Lemma 6.4 imply the existence of positive constants c'_τ, d'_τ dependent only on the ratio $r^*_\tau/r_{\tau 0}$ and on α_τ for which the inequality

$$\frac{2\pi}{n^*_\tau} \sum_{p=1}^{n^*_\tau} R\left(\left(\frac{r^k_{\tau\mu}}{r_{\tau 0}}\right)^{\lambda_\tau}, \lambda_\tau \theta^k_{\tau\mu}, \frac{(2p-1)\pi}{n^*_\tau}\right) \leq 1 + c'_\tau \exp\{-d'_\tau n\} \quad (6.9)$$

holds true.

If the left-hand side of inequality (6.8) is less than unity, then according to (4.7), $R^{km}_{\tau\mu} = R_\tau(r^k_{\tau\mu}, \theta^k_{\tau\mu}, \theta^m_\tau)$, $1 \leq m \leq n_\tau$, and, consequently, (6.7) is satisfied. Otherwise, according to (4.7) we have

$$\beta_\tau \sum_{m=1}^{n_\tau} R^{km}_{\tau\mu} = 1,$$

and, consequently, (6.7) is also valid by virtue of (6.8) and (6.9). We have proved Lemma 6.5.

Lemma 6.6. *If, for certain μ and k, $1 \leq \mu \leq M$, $1 \leq k \leq n_\mu$, the quantity $\tau = \tau(\mu, k)$ that can be found from (4.4) satisfies one of the two conditions*

$$1 \leq \tau \leq N, \qquad \nu_{\tau-1} + \nu_\tau \geq 1, \qquad (6.10)$$

$$N < \tau \leq L, \qquad \nu_{j(\tau)} = 1, \qquad (6.11)$$

then

$$\beta_\tau \sum_{m=1}^{n_\tau} R^{km}_{\tau\mu} \leq 1 + c'_\tau \exp\{-d'_\tau n\} - e_\tau, \qquad (6.12)$$

where c'_τ, d'_τ, e_τ are positive constants dependent only on the ratio $r^*_\tau/r_{\tau 0}$ and on α_τ.

Remark 6.1. Condition (6.10) means that a boundary condition of the first kind is defined at least on one side, $\gamma_{\tau-1}$ or γ_τ, of the angle in which the sector T_τ lies. Under condition (6.11) the boundary condition of the first kind is defined on the side $\gamma_{j(\tau)}$ of the polygon Ω which the half-disk T_τ adjoins.

Proof of Lemma 6.6. Suppose condition (6.11) is fulfilled. Then, in accordance with (3.12), (3.14), (3.16), (6.6) and with due account of the fact that $\alpha_\tau = 1$, $\beta_\tau = \alpha_\tau \pi / n_\tau = \pi / n_\tau$ we have

$$
\beta_\tau \sum_{m=1}^{n_\tau} R_{\tau\mu}^{km} \leq \beta_\tau \sum_{m=1}^{n_\tau} R_\tau(r_{\tau\mu}^k, \theta_{\tau\mu}^k, \theta_\tau^m)
$$

$$
= \frac{\pi}{n_\tau} \sum_{m=1}^{n_\tau} \left(R\left(\frac{r_{\tau\mu}^k}{r_{\tau 0}}, \theta_{\tau\mu}^k, \frac{(2m-1)\pi}{2n_\tau} \right) - R\left(\frac{r_{\tau\mu}^k}{r_{\tau 0}}, \theta_{\tau\mu}^k, \frac{(1-2m)\pi}{2n_\tau} \right) \right)
$$

$$
= \frac{2\pi}{2n_\tau} \sum_{m=1}^{2n_\tau} R\left(\frac{r_{\tau\mu}^k}{r_{\tau 0}}, \theta_{\tau\mu}^k, \frac{(2m-1)\pi}{2n_\tau} \right)
$$

$$
- \frac{2\pi}{n_\tau} \sum_{m=n_\tau+1}^{2n_\tau} R\left(\frac{r_{\tau\mu}^k}{r_{\tau 0}}, \theta_{\tau\mu}^k, \frac{(2m-1)\pi}{2n_\tau} \right). \tag{6.13}
$$

For τ being considered, the first sum on the right-hand side of (6.13) coincides with the sum appearing on the left-hand side of inequality (6.9) we have proved earlier and, by virtue of (3.14) and condition (4.4), the sum subtracted on the right-hand side of (6.13) is bounded from below by the quantity

$$
e_\tau = 2\pi \min_{0 \leq r \leq \rho_\tau} \min_{|\eta| < \infty} R(r, 0, \eta) = \frac{1 - \rho_\tau}{1 + \rho_\tau} > 0,
$$

where $\rho_\tau = r_\tau^* / r_{\tau 0} < 1$. This relation and (6.13) and (6.9) yield (6.12). Under condition (6.10) inequality (6.12) can be established by analogy. We have proved Lemma 6.6.

Lemma 6.7. *If one of the three conditions*

$$
1 \leq \tau \leq N, \qquad \nu_{\tau-1} = \nu_\tau = 0, \tag{6.14}
$$

$$
N < \tau \leq L, \qquad \nu_{j(\tau)} = 0, \tag{6.15}
$$

$$
L < \tau \leq M \tag{6.16}
$$

is satisfied by $\tau = \tau(\mu, k)$, *that can be found from formula (4.4), for certain* μ *and* k *such that* $1 \leq \mu \leq M$, $1 \leq k \leq n_\mu$, *then*

$$
R_{\tau\mu}^{km} \geq a_\tau', \tag{6.17}
$$

where $1 \le m \le n_\tau$, $a'_\tau > 0$ is a constant dependent only on the ratio $r^*_\tau / r_{\tau 0}$ and on α_τ.

Remark 6.2. Condition (6.14) means that a boundary condition of the second kind is defined on each of the two sides $\gamma_{\tau-1}$ and γ_τ of the angle in which the sector T_τ lies. Under condition (6.15) the boundary condition of the second kind is defined on the side $\gamma_{j(\tau)}$ which the half-disk T_τ adjoins. Under condition (6.16) the block T_τ is a disk.

Proof of Lemma 6.7. On the basis of (3.12), (3.14)–(3.18) we have

$$R_\tau(r^k_{\tau\mu}, \theta^k_{\tau\mu}, \theta^m_\tau) = \lambda_\tau(R((r^k_{\tau\mu}/r_{\tau 0})^{\lambda_\tau}, \lambda_\tau \theta^k_{\tau\mu}, \lambda_\tau \theta^m_\tau)$$
$$+ \xi_\tau R((r^k_{\tau\mu}/r_{\tau 0})^{\lambda_\tau}, \lambda_\tau \theta^k_{\tau\mu}, -\lambda_\tau \theta^m_\tau)),$$

where $\xi_\tau = 1$ and $\lambda_\tau = 1/\alpha_\tau$ under condition (6.14), $\xi_\tau = \lambda_\tau = \alpha_\tau = 1$ under condition (6.15) and, finally, $\xi_\tau = 0$, $\lambda_\tau = 1$, with $\alpha_\tau = 2$, under condition (6.16). Consequently, according to (3.14), (4.4),

$$R_\tau(r^k_{\tau\mu}, \theta^k_{\tau\mu}, \theta^m_\tau) \ge \frac{2}{\alpha_\tau} \min_{0 \le r \le \rho_\tau} \min_{|\eta| < \infty} R(r, 0, \eta) = \frac{1 - \rho_\tau}{\alpha_\tau \pi (1 + \rho_\tau)},$$

where $\rho_\tau = (r^*_\tau / r_{\tau 0})^{1/\alpha_\tau} < 1$. This result together with (4.7), (6.6), (6.8), and (6.9) yields (6.17) with constant

$$a'_\tau = (1 - \rho_\tau)/\alpha_\tau \pi (1 + \rho_\tau)(1 + c'_\tau) > 0$$

dependent only on the ratio $r^*_\tau / r_{\tau 0}$ and on α_τ.

Lemma 6.8. *If* $0 < \delta \le 1$, $0 \le a$, $0 \le b_1 \le b_2 \le \ldots \le b_M = b$, *with*

$$b_1 \le (1 - \delta)b + a, \quad b_t \le \delta b_{t-1} + (1 - \delta)b + a, \quad t = 2, 3, \ldots, M, \quad (6.18)$$

then

$$b \le Ma/\delta^M. \tag{6.19}$$

Proof. According to (6.18) we have

$$b \le \sum_{q=0}^{M-1} \delta^q((1 - \delta)b + a) \le (1 - \delta^M)b + Ma,$$

which implies (6.19). We have proved Lemma 6.8.

6.2.   Theorem on the Solvability of an Algebraic Problem

Consider the system of linear algebraic equations

$$w_\mu^k = a_\mu^k + \beta_\mu \sum_{m=1}^{n_\tau} w_\tau^m R_{\tau\mu}^{km}, \tag{6.20}$$

$$1 \leq \mu \leq M, \qquad 1 \leq k \leq n_\mu, \qquad \tau = \tau(\mu, k),$$

where $R_{\tau\mu}^{km}$ are coefficients (4.7) and a_μ^k are given numbers.

Theorem 6.1. *There exists a natural number n_0 such that for $n \geq n_0$ system (6.20) has a unique solution w_μ^k, $1 \leq \mu \leq M$, $1 \leq k \leq n_\mu$, with*

$$\|w\| \leq c\|a\|, \tag{6.21}$$

where $\| \cdot \|$ is norm (4.11) and c is a constant independent of $n \geq n_0$ and of $\|a\|$.

Proof.   Suppose that

$$\mu_1 = \min\{j \colon 1 \leq j \leq N, \, \nu_j = 1\}, \tag{6.22}$$

i.e., μ_1 is the smallest number of the side of polygon Ω on which a boundary condition of the first kind is defined,

$$\Omega_1 = \bigcup_{\substack{\mu=1 \\ \mu \neq \mu_1}}^{M} T_\mu, \qquad \Gamma_1 = \Omega \cap \overline{\Omega}_1 \setminus \Omega_1.$$

The set Ω_1, which is the union of a finite number of open extended blocks (sectors of disks, half-disks, and disks) lying in the polygon Ω, does not entirely include the extended sector T_{μ_1}, whose lateral sides lie on the sides γ_{μ_1-1} and γ_{μ_1} of the polygon Ω, since each one of the extended blocks T_μ, $\mu \neq \mu_1$, either does not intersect the sector T_{μ_1} or, according to conditions II-III (Sec. 2), is separated from the vertex P_{μ_1} by a positive distance. Therefore Ω_1 does not completely cover Ω. Consequently, $\Gamma_1 \neq \varnothing$. Since two noncoincident circles cannot intersect at more than two points, the set Γ_1 is the union of a finite number of circular arcs belonging to the boundaries of the extended blocks T_μ, $\mu \neq \mu_1$.

Since $r_\mu^* < r_{\mu 0}$, it follows that

$$\overline{T}_\mu^* \cap \Omega \subset T_\mu, \qquad 1 \leq \mu \leq M. \tag{6.23}$$

Therefore

$$\bigcup_{\mu \neq \mu_1} \overline{T}_\mu^* \cap \Omega \subset \Omega_1,$$

and, consequently, in accordance with condition IV (Sec. 2)

$$\Gamma_1 \subset \overline{T}_{\mu_1}^* \cap \Omega \setminus \bigcup_{\mu \neq \mu_1} \overline{T}_\mu^*. \tag{6.24}$$

We choose μ_2 such that the set $\Gamma_1 \cap \overline{T}_{\mu_2}$ contains an arc σ_2 of length $l_2 > 0$ of a circle of radius $r_{\mu_2 0}$. Obviously $\mu_2 \neq \mu_1$.

We set $\mathfrak{M}_q = \{\mu_i\}_{i=1}^q$, where $1 \leq \mu_i \leq M$, $\mu_i \neq \mu_k$ for $i \neq k$. The method of forming the sets $\mathfrak{M}_1$ and $\mathfrak{M}_2$ was indicated above. We give a recurrence method of constructing the sets $\mathfrak{M}_m$ for $m = 3, 4, \ldots, M$. Suppose the sets $\mathfrak{M}_m$ have been formed for $m = 1, 2, \ldots, q < M$. We set

$$\Omega_q = \bigcup_{\mu \notin \mathfrak{M}_q} T_\mu, \qquad \Gamma_q = \Omega \cap \overline{\Omega}_q \setminus \Omega_q.$$

Since $\Omega_q \neq \varnothing$, $\Omega_q \subset \Omega_1 \subset \Omega$, we have $\Gamma_q \neq \varnothing$. It follows that Γ_q is, like the set Γ_1, the union of a finite number of circular arcs belonging to the boundaries of the extended blocks T_μ, $\mu \notin \mathfrak{M}_q$. According to (6.23)

$$\bigcup_{\mu \notin \mathfrak{M}_q} \overline{T}_\mu^* \cap \Omega \subset \Omega_q,$$

and, consequently, by condition IV (Sec. 2)

$$\Gamma_q \subset \bigcup_{\mu \in \mathfrak{M}_q} \overline{T}_\mu^* \cap \Omega \setminus \bigcup_{\nu \notin \mathfrak{M}_q} \overline{T}_\nu^*. \tag{6.25}$$

By virtue of what has been said, we can choose $\mu_{q+1} \notin \mathfrak{M}_q$ such that the set $\Gamma_q \cap \overline{T}_{\mu_{q+1}}$ contains an arc σ_{q+1} of length[5] $l_{q+1} > 0$ of a circle (of radius $r_{\mu_{q+1} 0}$). We have thus formed the set $\mathfrak{M}_{q+1} = \{\mu_i\}_{i=1}^{q+1} = \mathfrak{M}_q \cup \mu_{q+1}$, and so on.

---

[5] For uniqueness it is possible, for instance, to take as μ_{q+1} the minimal $\mu \notin \mathfrak{M}_q$ such that $\Gamma_q \cap \overline{T}_\mu$ contains a circular arc of positive length.

Let

$$\mathfrak{R}_t = \{(\mu, k) : 1 \leq \mu \leq M, \, 1 \leq k \leq n_\mu, \, \tau(\mu, k) = \mu_t\}.$$

Since $n_\mu = \max\{4, [\alpha_\mu n]\}$, $\beta_\mu = \alpha_\mu \pi / n_\mu$, $1 \leq \mu \leq M$, there evidently exist positive constants n_0', c_{0t}, $2 \leq t \leq M$, such that for $n \geq n_0'$ the arc σ_t of positive length l_t contains

$$N_t \geq c_{0t} n_{\mu_t} \geq 1 \tag{6.26}$$

points of the set $\{P_{\mu_t}^m\}_{m=1}^{n_{\mu_t}}$.

If one of the three conditions

$$1 \leq \mu_t \leq N, \qquad \nu_{\mu_t - 1} = \nu_{\mu_t} = 0, \tag{6.27}$$

$$N < \mu_t \leq L, \qquad \nu_{j(\mu_t)} = 0, \tag{6.28}$$

$$L < \mu_t \leq M \tag{6.29}$$

is satisfied, then, for $(\mu, k) \in \mathfrak{R}_t$ we deduce from (6.26), the obvious inequality

$$\beta_{\mu_t} \sum_{m : P_{\mu_t}^m \in \sigma_t} R_{\mu_t \mu}^{km} \geq \beta_{\mu_t} N_t \min_{1 \leq m \leq n_{\mu_t}} R_{\mu_t \mu}^{km},$$

the equality $\beta_{\mu_t} = \alpha_{\mu_t} \pi / n_{\mu_t}$ and Lemma 6.7 that the inequality

$$\beta_{\mu_t} \sum_{m : P_{\mu_t}^m \in \sigma_t} R_{\mu_t \mu}^{km} \geq c^*, \qquad (\mu, k) \in \mathfrak{R}_t, \tag{6.30}$$

holds true for $n \geq n_0'$. Here $c^* = \pi \min_t \alpha_{\mu_t} c_{0t} a_{\mu_t}' > 0$ with the minimum being taken over the t such that one of the three conditions (6.27), (6.28), (6.29) is fulfilled.

If none of conditions (6.27), (6.28), (6.29) is satisfied for a certain t, $1 \leq t \leq M$, then, evidently, one of the two conditions

$$1 \leq \mu_t \leq N, \qquad \nu_{\mu_t - 1} + \nu_{\mu_t} \geq 1, \tag{6.31}$$

$$N < \mu_t \leq L, \qquad \nu_{j(\mu_t)} = 1 \tag{6.32}$$

must be fulfilled.

It follows from Lemma 6.6 that there are positive constants n_0'' and c^{**} such that for $n \geq n_0''$ each of the conditions (6.31), (6.32) implies the inequality

$$\beta_{\mu_t} \sum_{m=1}^{n_{\mu_t}} R_{\mu_t\mu}^{km} \leq 1 - c^{**}, \qquad (\mu, k) \in \mathfrak{R}_t. \qquad (6.33)$$

Finally, by virtue of (4.7) and (6.6) we have

$$0 \leq \beta_\tau \sum_{m=1}^{n_\tau} R_{\tau\mu}^{km} \leq 1, \qquad (6.34)$$

$$1 \leq \mu \leq M, \qquad 1 \leq k \leq n_\mu, \qquad \tau = \tau(\mu, k).$$

We set

$$n_0 = \max\{n_0', n_0''\}, \qquad (6.35)$$

$$\delta = \min\{c^*, c^{**}\}. \qquad (6.36)$$

Clearly, $0 < \delta \leq 1$ which follows from the positiveness of c^* and c^{**} and from (6.30), (6.33), (6.34).

We assume that system (6.20) has a solution w_μ^k, $1 \leq \mu \leq M$, $1 \leq k \leq n_\mu$, for some $n \geq n_0$. We set

$$W_t = \max\{|w_\mu^k| : (\mu, k) \in \bigcup_{p=1}^{t} \mathfrak{R}_p\}.$$

The set $\mathfrak{R}_1$ is nonempty for $n \geq n_0$ since by construction the arc σ_2 contains some points of the set $\{P_{\mu_2}^m\}_{m=1}^{n_{\mu_2}}$ whose number is positive according to (6.26). Therefore the obvious inequalities

$$0 \leq W_1 \leq W_2 \leq \ldots \leq W_M = \|w\| \qquad (6.37)$$

have sense.

Since condition (6.31) and, hence, inequality (6.33) are satisfied for $t = 1$ according to (6.22) and one of the conditions (6.27), (6.28), (6.29), (6.31), (6.32) is fulfilled for $2 \leq t \leq M$, it follows from $\sigma_t \subset \Gamma_{t-1}$, $t = 2, 3, \ldots, M$, on the basis of (6.20), (6.24), (6.25), (6.30), (6.33), (6.34), (6.36), (6.37) that

$$W_1 \leq (1 - \delta)\|w\| + \|a\|,$$

$$W_t \leq (1 - \delta)\|w\| + \delta W_{t-1} + \|a\|, \qquad t = 2, 3, \ldots, M.$$

It follows, according to (6.37) and Lemma 6.8, that the conjectured solution of system (6.20) satisfies inequality (6.21) for $c = M/\delta^M$. Thus, for $n \geq n_0$, where n_0 is defined by expression (6.35), the homogeneous system of linear algebraic equations corresponding to system (6.20) has only a trivial solution. Consequently, for any free terms and $n \geq n_0$, system (6.20) has a unique solution satisfying inequality (6.21), where $c = M/\delta^M$ does not depend on $n \geq n_0$ and $\|a\|$. We have proved Theorem 6.1.

Corollary 6.1. *There is a constant c independent of n and $\|a\|$ such that inequality (6.21) is satisfied for all n for which system (6.20) is uniquely solvable.*

Proof. If system (6.20) is uniquely solvable for some fixed n, then the existence of the constant $c = c_n$ independent of $\|a\|$ (and dependent, in general, on n), for which the solution of system (6.20) satisfies inequality (6.21), follows, for instance, from Cramer's rule for solving a system of linear algebraic equations. The corollary therefore follows from this result and from Theorem 6.1 since the system (6.20) is uniquely solvable only for a finite number of values $n < n_0$, where n_0 is quantity (6.35).

Corollary 6.2. *Lemma 4.1 holds true.*

6.3.   Fundamental Lemmas

Lemma 6.9. *Suppose $f_3(x) = f_1(x)f_2(x)$,*

$$\max_{x \in [\alpha, \beta]} |f_m^{(k)}(x)| \leq a_m^k k!, \qquad k = 0, 1, \ldots, \tag{6.38}$$

where a_m does not depend on k, $m = 1, 2$. Then inequality (6.38) is satisfied for $m = 3$ and $a_3 = 2 \max\{a_1, a_2\}$.

Lemma 6.9 follows from Leibniz' formula.

Remark 6.3. Suppose that ζ_j^+ is an open half-plane on whose boundary lies the side γ_j of the polygon Ω, and the inner normal to γ_j introduced earlier (Sec. 1) is directed toward this half-plane; $\gamma_j^0 = \gamma_j \setminus (P_j \cup P_{j+1})$, i.e., γ_j^0 is the side γ_j without endpoints; Ω_j^+ is the maximal open connected set of points belonging to the intersection of the open polygon Ω and the half-plane ζ_j^+, whose boundary contains the side γ_j^0; Ω_j^- is an open set obtained from Ω_j^+ by a mirror reflection with respect to the boundary of the half-plane ζ_j^+. We denote by Q_j^* a fixed harmonic polynomial which satisfies the boundary condition (1.2) on γ_j. This polynomial can be defined by analogy with (3.10). Suppose that P is a running point on $\Omega_j^+ \cup \gamma_j^0 \cup \Omega_j^-$, and $-P$ is a point which is a mirror image of the point P with respect to the boundary of the half-plane ζ_j^+. Then, evidently, the function

$$U_j(P) = \begin{cases} u(P), & P \in \Omega_j^+ \cup \gamma_j^0, \\ (-1)^{\nu_j}(u(-P) - Q_j^*(-P)) + Q_j^*(P), & P \in \Omega_j^-, \end{cases}$$

$$\text{(6.39)}$$

where u is a solution of the boundary-value problem (1.1), (1.2), is harmonic on $\Omega_j^+ \cup \gamma_j^0 \cup \Omega_j^-$, $1 \le j \le N$.

Lemma 6.10. *The following inequality is satisfied:*

$$\left| \frac{\partial^k u_\tau(r_{\tau 0}, \eta)}{\partial \eta^k} \right| \le a_\tau b_\tau^k k!, \qquad \text{(6.40)}$$

where u_τ is function (3.32), $0 \le \eta \le \alpha_\tau \pi$, $k = 0, 1, \ldots$, a_τ and b_τ are constants independent of k, $1 \le \tau \le M$.

Proof. By virtue of (3.1)–(3.11), (3.32), and conditions I–III (Sec. 2) imposed on the covering of the polygon Ω and Remark 6.3, the function u_τ admits of a harmonic extension from the extended block T_τ onto a sector (half-disk or disk) $T_\tau(r_\tau^0) \supset \overline{T}_\tau \cap \Omega$ (see (2.1)–(2.3)), where $r_\tau^0 > r_{\tau 0}$. Clearly, the quantity $r_\tau^0 > r_{\tau 0}$ can be such that the extended function, for which the designation u_τ is retained, is bounded on $T_\tau(r_\tau^0)$, $1 \le \tau \le M$, and, in addition, for $\tau \le L$, this function satisfies the homogeneous boundary conditions of the first or second kind on the rectilinear sides of the sector (half-disk) $T_\tau(r_\tau^0)$. Therefore the function u_τ can be oddly or evenly harmonically extended across these sides. Thus the set of points

$$\{(r_\tau, \eta) : r_\tau = r_{\tau 0}, 0 \le \eta \le \alpha_\tau \pi\}, \qquad 1 \le \tau \le M,$$

is strictly within a domain in which the function u_τ is harmonic and bounded. Applying a change of independent variables $\xi = \ln r_\tau$, $\eta = \eta$ (which retains the harmonicity) and Lemma 3.4, we arrive at inequality (6.40).

Lemma 6.11. *Let $r_\tau^{**} = (r_{\tau 0} + r_\tau^*)/2$. There holds an inequality*

$$\left| \int_0^{\alpha_\tau \pi} u_\tau(r_{\tau 0}, \eta) R_\tau(r_\tau, \theta_\tau, \eta) \, d\eta - \beta_\tau \sum_{m=1}^{n_\tau} u_\tau(r_{\tau 0}, \theta_\tau^m) R_\tau(r_\tau, \theta_\tau, \theta_\tau^m) \right|$$

$$\le c_\tau'' \exp\{-d_\tau'' n\}, \tag{6.41}$$

where $0 \le r_\tau \le r_\tau^{**}$, $0 \le \theta_\tau \le \alpha_\tau \pi$, u_τ is function (3.32), R_τ is the kernel defined by relations (3.12)–(3.18), c_τ'', $d_\tau'' > 0$ are constants independent of r_τ, θ_τ and of n, $1 \le \tau \le M$.

Proof. If $L < \tau \le M$, then $\alpha_\tau = 2$, $n_\tau \ge 2n$, $\theta_\tau^m = (2m - 1)\pi/n_\tau$, and, since the integrand function is 2π-periodic and infinitely differentiable with respect to η by virtue of (3.15), (3.14), (3.11), (3.32) and Remark 6.3, inequality (6.41) directly follows from Lemmas 6.3, 6.10, 6.9 and 6.1.

Let $N < \tau \le L$. Then $\alpha_\tau = 1$, $\theta_\tau^m = (2m - 1)\pi/2n_\tau$ and, by virtue of (3.16), (3.12)–(3.14), (3.10), (3.32)

$$\beta_\tau \sum_{m=1}^{n_\tau} u_\tau(r_{\tau 0}, \theta_\tau^m) R_\tau(r_\tau, \theta_\tau, \theta_\tau^m) - \int_0^{\alpha_\tau \pi} u_\tau(r_{\tau 0}, \eta) R_\tau(r_\tau, \theta_\tau, \eta) \, d\eta$$

$$= \frac{2\pi}{2n_\tau} \sum_{m=1}^{2n_\tau} u_\tau\left(r_{\tau 0}, \frac{(2m-1)2\pi}{2(2n_\tau)}\right) R\left(\frac{r_\tau}{r_{\tau 0}}, \theta_\tau, \frac{(2m-1)2\pi}{2(2n_\tau)}\right) -$$

$$- \int_0^{2\pi} u_\tau(r_{\tau 0}, \eta) R\left(\frac{r_\tau}{r_{\tau 0}}, \theta_\tau, \eta\right) d\eta, \tag{6.42}$$

where, for $\pi < \eta \le 2\pi$,

$$u_\tau(r_{\tau 0}, \eta) = (-1)^{\nu_{j(\tau)}} u_\tau(r_{\tau 0}, 2\pi - \eta), \tag{6.43}$$

in which $\nu_{j(\tau)} = 1$ ($\nu_{j(\tau)} = 0$), if a boundary condition of the first (second) kind is defined on the side of the polygon Ω which the half-disk T_τ adjoins by its diameter.

Clearly, the function $u_\tau(r_{\tau 0}, \eta)$ extended 2π-periodically with respect to η from the interval $[0, 2\pi]$, is infinitely differentiable on the η-axis. Because of the inequality $n_\tau \ge n$ and the fact that the function $R(r_\tau/r_{\tau 0}, \theta_\tau, \eta)$ is also 2π-periodic and infinitely differentiable with respect to η for $r_\tau \leq r_\tau^{**}$ (see (3.14)), and Lemmas 6.3, 6.10, 6.9, and 6.1, this, together with (6.42), (6.43), implies inequality (6.41) for $N < \tau \leq L$.

Suppose now that $1 \leq \tau \leq N$, $k_\tau = 4 - 2(\nu_{\tau-1}\nu_\tau + \bar{\nu}_{\tau-1}\bar{\nu}_\tau)$. Then $0 < \alpha_\tau \leq 2$, $\theta_\tau^m = (2m - 1)\alpha_\tau \pi/2n_\tau$ and, according to (3.12)–(3.14), (3.17), (3.18)

$$\int_0^{\alpha_\tau \pi} u_\tau(r_{\tau 0}, \eta) R_\tau(r_\tau, \theta_\tau, \eta) \, d\eta - \beta_\tau \sum_{m=1}^{n_\tau} u_\tau(r_{\tau 0}, \theta_\tau^m) R_\tau(r_\tau, \theta_\tau, \theta_\tau^m)$$

$$= \lambda_\tau \left(\int_0^{k_\tau \alpha_\tau \pi} u_\tau(r_{\tau 0}, \eta) R\left(\left(\frac{r_\tau}{r_{\tau 0}}\right)^{\lambda_\tau}, \lambda_\tau \theta_\tau, \lambda_\tau \eta \right) d\eta \right.$$

$$- \frac{k_\tau \alpha_\tau \pi}{k_\tau n_\tau} \sum_{m=1}^{k_\tau n_\tau} u_\tau \left(r_{\tau 0}, \frac{(2m - 1)k_\tau \alpha_\tau \pi}{2(k_\tau n_\tau)} \right)$$

$$\left. \times R\left(\left(\frac{r_\tau}{r_{\tau 0}}\right)^{\lambda_\tau}, \lambda_\tau \theta_\tau, \lambda_\tau \frac{(2m - 1)k_\tau \alpha_\tau \pi}{2(k_\tau n_\tau)} \right) \right), \tag{6.44}$$

where the function $u_\tau(r_{\tau 0}, \eta)$ is defined by expression (3.32) for $0 \leq \eta \leq \alpha_\tau \pi$,

$$u_\tau(r_{\tau 0}, \eta) = (-1)^{\nu_\tau - 1} u_\tau(r_{\tau 0}, 2\alpha_\tau \pi - \eta) \tag{6.45}$$

for $\alpha_\tau \pi < \theta_\tau \leq 2\alpha_\tau \pi$, and, finally,

$$u_\tau(r_{\tau 0}, \eta) = (-1)^{\nu_\tau} u_\tau(r_{\tau 0}, 4\alpha_\tau \pi - \eta) \tag{6.46}$$

for $k_\tau = 4$, $2\alpha_\tau \pi < \eta \leq 4\alpha_\tau \pi$.

We can establish by the method of analytic extension of the function u_τ (used in the proof of Lemma 6.10) that the function $u_\tau(r_{\tau 0}, \eta)$ has a period $k_\tau \alpha_\tau \pi$ with respect to η and is infinitely differentiable on the η-axis. Since $\lambda_\tau = 2/k_\tau \alpha_\tau$ (see (3.18)), the function $R((r_\tau/r_{\tau 0})^{\lambda_\tau}, \lambda_\tau \theta_\tau, \lambda_\tau \eta)$ also has a period $k_\tau \alpha_\tau \pi$ with respect to η and is infinitely differentiable with respect to η for $r_\tau \leq r_\tau^{**}$ (see (3.14)). Hence with due account of (6.44)–(6.46) (noting that $n_\tau \geq 4\alpha_\tau n/5$) and Lemmas 6.3, 6.10, 6.9, and 6.1 we have inequality (6.41) for $1 \leq \tau \leq N$.

We have proved Lemma 6.11.

Suppose that r, θ is a polar coordinate system chosen arbitrarily in the complex z-plane,

$$T^{**} = \{(r,\theta) : 0 < r < r^{**}, 0 < \theta < \alpha\pi\},$$

$$T^{*} = \{(r,\theta) : 0 < r < r^{*}, 0 < \theta < \alpha\pi\}$$

are sectors $(0 < r^{*} < r^{**}, 0 < \alpha \leq 2)$, P is their common vertex,

$$\gamma = \{(r,\theta) : r = r^{**}, 0 \leq \theta \leq \alpha\pi\},$$

$$\gamma^{0}_{2-k} = \{(r,\theta) : 0 < r < r^{**}, \theta = k\alpha\pi\}, \qquad k = 0,1.$$

Lemma 6.12. *Let w_ϵ be a solution of the boundary-value problem*

$$\Delta w_\epsilon = 0 \quad on \quad T^{**}, \qquad w_\epsilon = f_\epsilon(\theta) \quad on \quad \gamma, \qquad (6.47)$$

$$\nu_m w_\epsilon + \bar{\nu}_m (w_\epsilon)'_n = 0 \quad on \quad \gamma^0_m, \qquad m = 1,2, \qquad (6.48)$$

continuous up to the boundary of the sector T^{**}. Here $\nu_m = 0$ or 1, $\bar{\nu}_m = 1 - \nu_m$, $(w_\epsilon)'_n$ is the derivative of the function w_ϵ along the normal to γ^0_m, $\lambda = 1/(2-\nu_1\nu_2-\bar{\nu}_1\bar{\nu}_2)\alpha$, $f_\epsilon(\theta)$ is an arbitrary function continuous on the interval $[0,\alpha\pi]$ with

$$\max_{0 \leq \theta \leq \alpha\pi} |f_\epsilon(\theta)| \leq \epsilon \qquad (6.49)$$

and, in addition, if $\nu_1 = 1$, then $f_\epsilon(\alpha\pi) = 0$, and if $\nu_2 = 1$, then $f_\epsilon(0) = 0$. Then the inequality

$$\left| \frac{\partial^p w_\epsilon}{\partial x^{p-q} \partial y^q} \right| \leq c^0_p \epsilon \qquad (6.50)$$

is satisfied on $\overline{T}^{*}$ for integer λ and any ν_1 and ν_2 when $p \geq \lambda$, the inequality

$$\left| \frac{\partial^p w_\epsilon}{\partial x^{p-q} \partial y^q} \right| \leq c^0_p r^{\lambda-p} \epsilon \qquad (6.51)$$

is satisfied on $\overline{T}^{*}$ for any λ if $\nu_1 + \nu_2 \geq 1$, $0 \leq p < \lambda$ or $\nu_1 = \nu_2 = 0$, $1 \leq p < \lambda$, the inequality

$$\left| \frac{\partial^p w_\epsilon}{\partial x^{p-q} \partial y^q} \right| \leq c^0_p r^{\lambda-p} \epsilon \qquad (6.52)$$

is satisfied on $\overline{T}^* \setminus P$ for noninteger λ and any ν_1 and ν_2 when $p > \lambda$.

Everywhere $0 \le q \le p$, c_p^0 is a constant independent of $r = r(x, y)$, $\theta = \theta(x, y)$ and of ε.

Proof. We assume that λ is an integer. Let

$$K^{**} = \{(r, \theta) : 0 \le r < r^{**}, 0 \le \theta < 2\pi\}.$$

We harmonically extend the function w_ε from T^{**} onto $K^{**} \setminus P$ and retain the same notation for the extended function. The extension is carried out from the sector T^{**} across one of the sides to the adjacent sector of the angular magnitude $\alpha\pi$. The extension is odd (even) with respect to the chosen side if a homogeneous boundary condition of the first (second) kind is given on it in (6.48). Then the function is extended in the same way to the next sector of the same magnitude, etc. ($2/\alpha - 1$ times). At the point P, i.e., at the center of the disk K^{**}, the function w_ε has a removable singularity and, consequently, is harmonic in the whole disk K^{**}. Clearly, w_ε is continuous on $\overline{K}^{**}$. According to (6.49), by the principle of maximum, $|w_\varepsilon| \le \varepsilon$ on $\overline{K}^{**}$. Since the closed sector $\overline{T}^*$ lies strictly within the disk K^{**} at the distance of $\rho = r^{**} - r^* > 0$ from its boundary, we have, according to Lemma 3.4

$$\left| \frac{\partial^p w_\varepsilon}{\partial x^{p-q} \partial y^q} \right| \le \frac{p! 2^{p+2}}{\pi (r^{**} - r^*)^p} \varepsilon \quad \text{on} \quad \overline{T}^*,$$

where $p = 1, 2, \ldots, 0 \le q \le p$. Thus, in particular, we have proved inequality (6.50) for all the values of the parameters λ, ν_1, ν_2 and p corresponding to it.

Let us now prove inequalities (6.51) and (6.52) for $\nu_1 = \nu_2 = 1$ and the values of λ and p corresponding to them. Let $r' = (r^{**} + r^*)/2$,

$$T' = \{(r, \theta) : 0 < r < r', 0 < \theta < \alpha\pi\},$$

$$\gamma' = \{(r, \theta) : r = r', 0 \le \theta \le \alpha\pi\}.$$

By virtue of (6.48) we can extend the function w_ε oddly across γ_1^0 and γ_2^0 for $\nu_1 = \nu_2 = 1$ preserving the harmonicity. Consequently, any point of the arc γ' can serve as the center of a closed disk of some fixed radius $\rho' > 0$, the harmonic function extended in this way being defined on this disk. According to Lemma 3.4, with due account of (6.49) and the maximum principle, this result implies the existence of a constant a such that the inequality

$$\max_{s \in \gamma'} \left| \frac{\partial w_\varepsilon}{\partial s} \right| \leq a\varepsilon, \tag{6.53}$$

where s is the arc length reckoned along γ', holds true. Since the function w_ε vanishes at the endpoints of the arc γ' by the hypothesis, this function satisfies the inequality

$$|w_\varepsilon(r', \theta)| \leq ar'\varepsilon \cdot \begin{cases} \theta, & 0 \leq \theta \leq \alpha\pi/2, \\ \alpha\pi - \theta, & \alpha\pi/2 < \theta \leq \alpha\pi, \end{cases}$$

on γ' on the basis of (6.53).

If the homogeneous boundary conditions (6.48) (for $\nu_1 = \nu_2 = 1$) and the maximum principle are taken into account, it follows that the harmonic function

$$W(r, \theta) = b\varepsilon r^{1/\alpha} \sin \frac{\theta}{\alpha}, \tag{6.54}$$

where $b = a\alpha\pi(r')^{1-1/\alpha}/2$, is a majorant for the function w_ε on the closed sector $\overline{T}'$, i.e.,

$$|w_\varepsilon(r, \theta)| \leq W(r, \theta) \quad \text{on} \quad \overline{T}'. \tag{6.55}$$

Relations (6.54) and (6.55) directly yield inequality (6.51) for $p = 0$, $\nu_1 = \nu_2 = 1$.

Let us oddly, i.e., harmonically, extend the function w_ε from the sector T' across its sides. Placing the running point (r, θ), belonging to $\overline{T}^* \setminus P$, at the center of a disk of radius $c'r$ ($c' > 0$ is a constant) which entirely lies in the domain of definition of the extended function and, using Lemma 3.4 with due account of (6.54) and (6.55), we arrive at inequality (6.51) for $1 \leq p < \lambda$ and at inequality (6.52) for $p > \lambda$ on $\overline{T}^* \setminus P$ (for $\nu_1 = \nu_2 = 1$). By continuity, inequality (6.51) is also valid at the vertex P for the values of p pertaining to it.

For $\nu_1 + \nu_2 = 1$ inequalities (6.51) and (6.52) can be proved by analogy with the only difference that the function w_ε is first harmonically extended from T^{**} onto the sector with a twice as large vertex angle across the side on which a boundary condition of the second kind is given in (6.48). If $\alpha > 1$, then the doubled sector lies on a Riemannian surface.

For $\nu_1 = \nu_2 = 0$ we consider a function v_ε, conjugate to w_ε, which is harmonic on T^{**}. This function is continuous up to $\gamma_1^0 \cup \gamma_2^0 \cup P$. The continuity on $T^{**} \cup \gamma_1^0 \cup \gamma_2^0$ is obvious and the continuity of the function v_ε at the vertex P of the sector T^{**} follows from the results of [50] and the Cauchy-Riemann conditions according to which the order of growth of the derivatives $\partial v_\varepsilon / \partial x$ and $\partial v_\varepsilon / \partial y$ in the vicinity of the point P is not higher than that of $r^{1/\alpha - 1}$, where $1/\alpha \geq 1/2$.

Inasmuch as $(w_\varepsilon)'_n = 0$ on γ_m^0, $m = 1, 2$, it follows, by virtue of the Cauchy-Riemann conditions and the continuity of the function v_ε at the vertex P, that

$$v_\varepsilon = c \quad \text{on} \quad \gamma_1^0 \cup \gamma_2^0 \cup P,$$

where c is a constant. We fix the function v_ε such that the equality $c = 0$ is satisfied.

By analogy with (6.53) we find that

$$\left| \frac{\partial v_\varepsilon}{\partial s} \right| = \left| \frac{\partial w_\varepsilon}{\partial r} \right| \leq a\varepsilon \quad \text{on} \quad \gamma'.$$

Next we construct the estimates for the derivatives of all orders of the function v_ε as we did before for $\nu_1 = \nu_2 = 1$. Then, with the use of Cauchy-Riemann conditions, we obtain the estimates for the derivatives of the function w_ε in the same form (6.51), (6.52) for corresponding values of λ and p (for $\nu_1 = \nu_2 = 0$). The proof of Lemma 6.12 is completed.

6.5.   Proof of Theorem 5.1

Suppose that for a certain n the system of linear algebraic equations (4.8) has a unique solution u_μ^k, $1 \le \mu \le M$, $1 \le k \le n_\mu$. The unique solvability of system (4.8) is guaranteed by Theorem 6.1, at least for a sufficiently large n.

Let us first prove the inequality

$$\max_{1 \le \mu \le M} \max_{1 \le k \le n_\mu} |u_\mu^k - u(P_\mu^k)| \le c' \exp\{-d'n\}, \qquad (6.57)$$

where $u(P_\mu^k)$ is the value of the solution u of the boundary-value problem (1.1), (1.2) at the point P_μ^k (see Sec. 4); c', $d' > 0$ are some constants independent of n for which system (4.8) has a unique solution.

In accordance with (3.32), (4.6), (4.8), and Theorem 3.1, for $1 \le \mu \le M$, $1 \le k \le n_\mu$, $\tau = \tau(\mu, k)$ we have

$$u_\mu^k - u(P_\mu^k) = \beta_\tau \sum_{m=1}^{n_\tau} (u_\tau^m - u(P_\tau^m)) R_{\tau\mu}^{km}$$

$$+\beta_\tau \sum_{m=1}^{n_\tau} u_\tau(r_{\tau 0}, \theta_\tau^m)(R_{\tau\mu}^{km} - R_\tau(r_{\tau\mu}^k, \theta_{\tau\mu}^k, \theta_\tau^m))$$

$$+\beta_\tau \sum_{m=1}^{n_\tau} u_\tau(r_{\tau 0}, \theta_\tau^m) R_\tau(r_{\tau\mu}^k, \theta_{\tau\mu}^k, \theta_\tau^m)$$

$$-\int_0^{\alpha_\tau \pi} u_\tau(r_{\tau 0}, \eta) R_\tau(r_{\tau\mu}^k, \theta_{\tau\mu}^k, \eta) \, d\eta.$$

Since $r_{\tau\mu}^k \le r_\tau^* < r_\tau^{**} < r_{\tau 0}$, $1 \le \mu \le M$, $1 \le k \le n_\mu$, and the function $u_\tau(r_\tau, \theta_\tau)$ is bounded on $\overline{T}_\tau$, $1 \le \tau \le M$, relation (6.6), Lemmas 6.5, 6.11 and Corollary 6.1 of Theorem 6.1 yield inequality (6.56) with some positive constants c' and d' independent of n for which system (4.8) is uniquely solvable.

According to (3.32), (5.1), Remarks 3.2, 3.7, 5.1 and Theorem 3.1 we have, on the closed intermediate block $\overline{T}_\mu^{**} \supset \overline{T}_\mu^*$ (see (5.2)), $1 \le \mu \le M$,

$$u_{\mu n}(r_\mu, \theta_\mu) - u(r_\mu, \theta_\mu)$$

$$= \beta_\mu \sum_{k=1}^{n_\mu} u_\mu(r_{\mu 0}, \theta_\mu^k) R_\mu(r_\mu, \theta_\mu, \theta_\mu^k) - \int_0^{\alpha_\mu \pi} u_\mu(r_{\mu 0}, \eta) R_\mu(r_\mu, \theta_\mu, \eta)\, d\eta$$

$$+ \beta_\mu \sum_{k=1}^{n_\mu} (u_\mu^k - u(P_\mu^k)) R_\mu(r_\mu, \theta_\mu, \theta_\mu^k),$$

where $u(r_\mu, \theta_\mu)$ is a solution of the boundary-value problem (1.1), (1.2),

$$u(P_\mu^k) = u(r_{\mu 0}, \theta_\mu^k).$$

Hence, by virtue of Lemmas 6.3, 6.11 and estimate (6.56), we arrive at an inequality

$$|u_{\mu n}(r_\mu, \theta_\mu) - u(r_\mu, \theta_\mu)| \le c_0^* \exp\{-d_0 n\} \qquad (6.58)$$

on $\overline{T}_\mu^{**} \supset \overline{T}_\mu^*$, $1 \le \mu \le M$, where c_0^*, $d_0 > 0$ are constants independent of r_μ, θ_μ, of n for which system (4.8) is uniquely solvable, and of μ which assumes a fixed number of values.

Inequality (6.57), in particular, includes all cases from inequality (5.3) for $p = 0$ when $c_0 = c_0^*$. In addition, relation (6.57) and Lemma 3.4 yield inequality (5.3) for $p \ge 1$ when $L < \mu \le M$, since $\overline{T}_\mu^* \subset T_\mu^{**}$. For all the other cases inequality (5.3) as well as inequalities (5.4) and (5.5) follow from inequality (6.57) and Lemma 6.12 for all values of the parameters corresponding to them. This completes the proof of Theorem 5.1.

6.6. Proofs of Lemmas 4.2–4.4

Let us first prove an auxiliary lemma.

Lemma 6.13. *Suppose that the numbers $z_{0\mu}^k$ are arbitrary, $\varepsilon_{\nu\mu}^k$ are quantities defined by relations (4.9) and (4.10), and*

$$z_{\nu\mu}^k = \beta_\tau \sum_{m=1}^{n_\tau} z_{\nu-1,\tau}^m R_{\tau\mu}^{km}, \qquad \nu = 1, 2, \ldots,$$

where $R_{\tau\mu}^{km}$ are coefficients (4.7), $\tau = \tau(\mu, k)$ is defined by formula (4.4), $1 \le \mu \le M$, $1 \le k \le n_\mu$. Then the inequality

$$\|z_\nu\| \le \|\varepsilon_\nu\| \cdot \|z_0\|, \qquad \nu = 0, 1, \ldots, \qquad (6.59)$$

where $\|\cdot\|$ *is norm* (4.11), *holds true.*

Proof.   We set $\bar{z}_{0\mu}^k = |z_{0\mu}^k|$,

$$\bar{z}_{\nu\mu}^k = \beta_\tau \sum_{m=1}^{n_\tau} \bar{z}_{\nu-1,\tau}^m R_{\tau\mu}^{km},$$

$1 \le \mu \le M, 1 \le k \le n_\mu, \tau = \tau(\mu, k), \nu = 1, 2, \ldots.$

By virtue of (6.6) it is evident that

$$\|z_\nu\| \le \|\bar{z}_\nu\|, \qquad \nu = 0, 1, \ldots. \tag{6.60}$$

Suppose the inequalities

$$0 \le \bar{z}_{\nu\mu}^k \le \varepsilon_{\nu\mu}^k \|z_0\| \tag{6.61}$$

hold for $1 \le \mu \le M, 1 \le k \le n_\mu, 0 \le \nu \le p$, where p is a nonnegative integer. Then, according to (4.10), (6.6), (6.60), we have

$$0 \le \bar{z}_{p+1,\mu}^k = \beta_\tau \sum_{m=1}^{n_\tau} \bar{z}_{p\tau}^m R_{\tau\mu}^{km} \le \|z_0\| \beta_\tau \sum_{m=1}^{n_\tau} \varepsilon_{p\tau}^m R_{\tau\mu}^{km} = \varepsilon_{p+1,\mu}^k \|z_0\|$$

for $1 \le \mu \le M, 1 \le k \le n_\mu, \tau = \tau(\mu, k)$.

Since conditions (6.60) are trivially satisfied for $p = 0$ (see (4.9)), their fulfillment for all nonnegative integers ν follows by induction from what has been proved, i.e., in particular, the inequality $\|\bar{z}_\nu\| \le \|\varepsilon_\nu\| \cdot \|z_0\|$ holds true, and, by virtue of (6.59), inequality (6.58) is valid for $\nu = 0, 1, \ldots.$

Proof of Lemma 4.2. It can be seen from (6.6) and (6.34) that the quantities $\varepsilon_{\nu\mu}^k$, defined by relations (4.9) and (4.10), satisfy the inequalities

$$0 \le \varepsilon_{\nu+1,\mu}^k \le \varepsilon_{\nu\mu}^k \le 1, \tag{6.62}$$

for $1 \le \mu \le M, 1 \le k \le n_\mu, \nu = 0, 1, \ldots$, which imply the existence of limits

$$\lim_{\nu \to \infty} \varepsilon_{\nu\mu}^k = \bar{\varepsilon}_\mu^k \ge 0, \qquad 1 \le \mu \le M, \qquad 1 \le k \le n_\mu.$$

Passing to the limit as $\nu \to \infty$ in equalities (4.10), we see that $\bar{\varepsilon}_\mu^k$, $1 \le \mu \le M, 1 \le k \le n_\mu$, is a solution of the homogeneous system of linear algebraic equations corresponding to system (4.8).

We assume that the right-hand inequality in (6.34) becomes an equality for all μ and k. Then evidently, $\varepsilon_{\nu\mu}^k = 1$, $1 \le \mu \le M$, $1 \le k \le n_\mu$, $\nu = 0, 1, \ldots$, and

$$\|\varepsilon_{M+1}\| = \|\tilde{\varepsilon}\| = 1,$$

i.e., the homogeneous system corresponding to (4.8) has a nontrivial solution. Thus, in the case being considered condition (4.12) is not fulfilled and system (4.8) is not uniquely solvable.

Suppose now that

$$\min_{1 \le \mu \le M} \min_{1 \le k \le n_\mu} \beta_\tau \sum_{m=1}^{n_r} R_{\tau\mu}^{km} < 1, \tag{6.63}$$

where $\tau = \tau(\mu, k)$. To begin with we assume that

$$\|\varepsilon_{M+1}\| = 1. \tag{6.64}$$

We introduce the notation

$$A_\nu = \{\mu : 1 \le \mu \le M, \ \min_{1 \le k \le n_\mu} \varepsilon_{\nu\mu}^k = 1\},$$

and let p_ν be the number of elements of the set A_ν. On the basis of (4.9) and (6.61) we have

$$A_0 \supseteq A_1 \supseteq A_2 \supseteq \cdots,$$
$$M = p_0 \ge p_1 \ge p_2 \ge \cdots \tag{6.65}$$

By virtue of (4.10), (6.34), and (6.61), equality (6.63) means that $A_M \ne \varnothing$, i.e., $p_M \ge 1$, and condition (6.62) implies that $A_1 \ne A_0$, i.e., $p_1 < M$. It follows from these results and from (6.64) that there is a ν^*, $1 \le \nu^* \le M - 1$, such that

$$A_{\nu^*} = A_{\nu^*+1} \ne \varnothing. \tag{6.66}$$

Condition (6.65) together with (6.34) and (6.61), means that if $\mu \in A_{\nu^*}$, then we have $\tau = \tau(\mu, k) \in A_{\nu^*}$ for any k, $1 \le k \le n_\mu$, with

$$\beta_\tau \sum_{m=1}^{n_r} R_{\tau\mu}^{km} = 1. \tag{6.67}$$

But then, evidently, if $\mu \in A_{\nu^\bullet+1}$, i.e., $\mu \in A_{\nu^\bullet}$, then $\tau = \tau(\mu, k) \in A_{\nu^\bullet}$ for any k, $1 \leq k \leq n_\mu$, i.e., $\tau \in A_{\nu^\bullet+1}$, with (6.66) being satisfied. Therefore, according to (4.10), (6.66), for any $\mu \in A_{\nu^\bullet+1}$ we have

$$\varepsilon^k_{\nu^\bullet+2,\mu} = \beta_\tau \sum_{m=1}^{n_\tau} \varepsilon^m_{\nu^\bullet+1,\tau} R^{km}_{\tau\mu} = \beta_\tau \sum_{m=1}^{n_\tau} R^{km}_{\tau\mu} = 1, \qquad (6.68)$$

$$1 \leq k \leq n_\mu,$$

where $\tau = \tau(\mu, k) \in A_{\nu^\bullet+1}$, $\varepsilon^m_{\nu^\bullet+1,\tau} = 1$, $1 \leq m \leq n_\tau$.

Thus $A_{\nu^\bullet+2} = A_{\nu^\bullet+1} = A_{\nu^\bullet} \neq \varnothing$ by virtue of (6.64), (6.65), and (6.67). Analogously, by induction we infer that

$$A_{\nu^\bullet+p} = A_{\nu^\bullet} \neq \varnothing, \qquad p = 1, 2, \ldots. \qquad (6.69)$$

But (6.68) means that $\|\tilde{\varepsilon}\| = 1$, i.e., the homogeneous system corresponding to system (4.8) has a nontrivial solution.

Thus, whenever conditions (4.12) are not fulfilled, the system of linear algebraic equations (4.8) is not uniquely solvable.

We assume, finally, that condition (4.12) is fulfilled. Suppose the homogeneous system of linear algebraic equations corresponding to system (4.8) has a solution $\tilde{z}^k_\mu$, $1 \leq \mu \leq M$, $1 \leq k \leq n_\mu$. Since

$$\tilde{z}^k_\mu = \beta_\tau \sum_{m=1}^{n_\tau} \tilde{z}^m_\tau R^{km}_{\tau\mu}, \qquad 1 \leq \mu \leq M, \qquad 1 \leq k \leq n_\mu,$$

we can set $z^k_{\nu\mu} = \tilde{z}^k_\mu$, $\nu = 0, 1, \ldots$, and arrive, on the basis of Lemma 6.13, at an inequality $\|\tilde{z}\| \leq \|\varepsilon_{M+1}\| \cdot \|\tilde{z}\|$, which, by virtue of (4.12), is possible only for $\|\tilde{z}\| = 0$.

Consequently, system (4.8) is uniquely solvable under condition (4.12). We have proved Lemma 4.2.

Proof of Lemma 4.3. ‖The proof of this lemma is closely connected with the proof of Theorem 6.1. Let $n \geq n_0$, where n_0 is quantity (6.35). Then system (4.8) is uniquely solvable.

We set

$$E_{\nu m} = \max_{1 \leq t \leq m} \max_{(\mu,k) \in \mathfrak{R}_t} \varepsilon^k_{\nu\mu}, \qquad (6.70)$$

where $1 \leq m \leq M$, $\nu = 0, 1, \ldots$, $\varepsilon_{\nu\mu}^k$ are quantities defined by relations (4.9), (4.10), $\mathfrak{R}_t$ is a set defined in the proof of Theorem 6.1, with $\mathfrak{R}_1 \neq \varnothing$. Since

$$\bigcup_{t=1}^{M} \mathfrak{R}_t = \bigcup_{\mu=1}^{M} \bigcup_{k=1}^{n_\mu} (\mu, k),$$

it follows by virtue of (4.11) and (6.69) that

$$E_{\nu M} = \|\varepsilon_\nu\|, \qquad \nu = 0, 1, \ldots. \tag{6.71}$$

For $t = 1$ condition (6.31) and, hence, (6.33) are satisfied according to (6.22). Therefore

$$E_{11} \leq (1 - \delta)\|\varepsilon_0\|, \tag{6.72}$$

where $\delta \in (0, 1)$ is quantity (6.36).

Inasmuch as one of the conditions (6.30) or (6.33) is satisfied for $2 \leq t \leq M$, it follows that

$$E_{tt} \leq (1 - \delta)\|\varepsilon_0\| + \delta E_{t-1,t-1}, \qquad t = 2, 3, \ldots, M, \tag{6.73}$$

by virtue of (4.10), (6.6), (6.34), (6.61) and the inclusions $\sigma_t \subset \Gamma_{t-1}$, $t = 2, 3, \ldots, M$.

Since $\|\varepsilon_0\| = 1$, we easily get from (6.70)–(6.72) an inequality

$$\|\varepsilon_M\| \leq 1 - \delta^M,$$

in which the right-hand side $1 - \delta^M < 1$ does not depend on $n \geq n_0$. Since $\|\varepsilon_{M+1}\| \leq \|\varepsilon_M\|$ in accordance with (6.61), a fortiori

$$\|\varepsilon_{M+1}\| \leq 1 - \delta^M < 1, \qquad n \geq n_0, \tag{6.74}$$

where n_0 is quantity (6.35).

System (4.8) can be uniquely solvable only for a finite number of natural numbers n which are smaller than the indicated n_0. Therefore (6.73) and Lemma 4.2 yield the statement of Lemma 4.3.

Proof of Lemma 4.4. Inequality (4.16) follows from (6.61). Let us prove the left-hand inequality in (4.17). We assume that it is satisfied for $\nu = \nu(p) = p(M+1)$, where p is a natural number. Then, setting $z_{0\mu}^k = \varepsilon_{\nu(p)\mu}^k$, $1 \leq \mu \leq M$, $1 \leq k \leq n_\mu$, we obtain

$$\|\varepsilon_{\nu(p)+M+1}\| = \|z_{M+1}\| \leq \|\varepsilon_{M+1}\| \cdot \|z_0\| = \|\varepsilon_{M+1}\| \cdot \|\varepsilon_{\nu(p)}\| \leq \|\varepsilon_{M+1}\|^{p+1}$$

on the basis of Lemma 6.13.

Since the left-hand inequality in (4.17) is trivially satisfied for $\nu = M+1$, i.e., for $p = 1$, it follows by induction from what has been proved that the left-hand inequality in (4.17) is satisfied for all ν of the form $\nu = p(M+1)$, where p is a natural number. For the other values of ν the left-hand inequality of (4.17) follows from the case proved above and from (4.16). The right-hand inequality in (4.17) is obvious.

Suppose now that $z_{0\mu}^k = u_{0\mu}^k - u_\mu^k = -u_\mu^k$, $1 \leq \mu \leq M$, $1 \leq k \leq n_\mu$. Then according to (4.8) and (4.15), Lemma 6.13 yields

$$\|u_\nu - u\| = \|z_\nu\| \leq \|\varepsilon_\nu\| \cdot \|z_0\| = \|\varepsilon_\nu\| \cdot \|u\|, \qquad \nu = 0, 1 \ldots, \quad (6.75)$$

where u is a solution of system (4.8). Hence

$$\|u\| \leq \|u - u_\nu\| + \|u_\nu\| \leq \|\varepsilon_\nu\| \cdot \|u\| + \|u_\nu\|,$$

and consequently, by virtue of (4.12), (4.16), and (6.74),

$$\|u_\nu - u\| \leq \frac{\|\varepsilon_\nu\|}{1 - \|\varepsilon_\nu\|} \|u_\nu\| \tag{6.76}$$

for $\nu \geq M+1$.

In accordance with (4.16) and Lemma 4.3 we have

$$\frac{1}{1 - \|\varepsilon_\nu\|} \leq \frac{1}{1 - \|\varepsilon_{M+1}\|} \leq \frac{1}{1 - \sigma_0} = c_1, \tag{6.77}$$

where $\nu \geq M+1$, c_1 does not depend on ν and n. On the basis of (6.74) and (6.61) we get

$$\|u_\nu\| \leq \|u_\nu - u\| + \|u\| \leq (\|\varepsilon_\nu\| + 1)\|u\| \leq 2\|u\| \tag{6.78}$$

for $\nu \geq 0$.

Inasmuch as the carrier function $Q_\mu(r_\mu, \theta_\mu)$, defined by (3.2), (3.4), (3.6), (3.9), (3.10) or (3.11), is bounded on the closed basic block $\overline{T}_\mu$, $1 \le \mu \le M$, it follows, by virtue of (6.6), (6.34) that the free terms of system (4.8), expressed by formula (4.14), satisfy the inequality $\|a\| \le c'$, where c' does not depend on $n \ge 1$. Consequently, by virtue of Lemma 4.2 and Corollary 6.1 of Theorem 6.1, the inequality

$$\|u\| \le c\|a\| \le cc' = c_2, \tag{6.79}$$

where c_2 is independent of n, holds for the solution of system (4.8) under condition (4.12).

Under condition (4.12), inequalities (6.75)–(6.78) imply (4.18), where $c_* = 2c_1 c_2$ does not depend on ν and n. Lemma 4.4 is proved.

Remark 6.4. In some inequalities (Lemmas 6.5–6.7, 6.11), which we use to prove Theorem 6.1 and the fundamental Theorem 5.1, the constants depend on the ratios $r_\mu^* / r_{\mu 0}$ of the radii of the basic and extended block-disks, half-disks, sectors of disks, $1 \le \mu \le M$. When the ratio $r_\mu^* / r_{\mu 0}$ tends to unity, the corresponding constants tend to infinity (Lemmas 6.3, 6.4) and for $r_\mu^* / r_{\mu 0} = 1$ the indicated inequalities have no sense in the general case. Therefore the requirement that the inequalities $r_\mu^* < r_{\mu 0}$, $1 \le \mu \le M$ be satisfied, that we introduced in Sec. 2 when constructing the covering of the polygon Ω by pairs of geometrically similar blocks, is essentially defined. We can extend the concept of a basic block T_μ^* considering the basic block to be a connected part (subdomain) of the extended block T_μ separated by a positive distance from the boundary of the extended block over which the integration in (3.31) is carried out for the required solution u. The fundamental Theorem 5.1 remains valid when condition IV (Sec. 2) imposed on the covering of the polygon Ω is preserved.

7. The Stability and the Labor Content of Computations Required by the Block Method

7.1. Preliminary Lemmas

Suppose

$$\tilde{R}_{\tau\mu}^{km} \ge 0, \tag{7.1}$$

$$1 \leq \mu \leq M, \quad 1 \leq k \leq n_\mu, \quad \tau = \tau(\mu, k), \quad 1 \leq m \leq n_\tau,$$

are arbitrary numbers, the inequality

$$\beta_\tau \sum_{m=1}^{n_\tau} \tilde{R}_{\tau\mu}^{km} \leq 1 \tag{7.2}$$

is satisfied for all indicated μ and k ($\tau(\mu, k)$ is determined from formula (4.4) and n_τ and β_τ from formulas (4.1) and (4.2)).

We set

$$\tilde{\varepsilon}_{0\mu}^k = 1, \tag{7.3}$$

$$\tilde{\varepsilon}_{\nu\mu}^k = \beta_\tau \sum_{m=1}^{n_\tau} \tilde{\varepsilon}_{\nu-1,\tau}^m \tilde{R}_{\tau\mu}^{km}, \quad \nu = 1, 2, \ldots, \tag{7.4}$$

$$1 \leq \mu \leq M, \quad 1 \leq k \leq n_\mu, \quad \tau = \tau(\mu, k).$$

Lemma 7.1. *Suppose that the numbers $\tilde{z}_{0\mu}^k$ are arbitrary and*

$$\tilde{z}_{\nu\mu}^k = \beta_\tau \sum_{m=1}^{n_\tau} \tilde{z}_{\nu-1,\tau}^m \tilde{R}_{\tau\mu}^{km}, \quad \nu = 1, 2, \ldots, \tag{7.5}$$

where $1 \leq \mu \leq M$, $1 \leq k \leq n_\mu$, $\tau = \tau(\mu, k)$. Then the inequality

$$\|\tilde{z}_\nu\| \leq \|\tilde{\varepsilon}_\nu\| \cdot \|\tilde{z}_0\|, \quad \nu = 0, 1 \ldots,$$

where $\| \cdot \|$ is norm (4.11), is satisfied.

Lemma 7.1 can be proved by analogy with Lemma 6.13.

Let us consider a system of linear algebraic equations for the unknowns w_μ^k:

$$w_\mu^k = b_\mu^k + \beta_\tau \sum_{m=1}^{n_\tau} w_\tau^m \tilde{R}_{\tau\mu}^{km}, \tag{7.6}$$

$$1 \leq \mu \leq M, \quad 1 \leq k \leq n_\mu, \quad \tau = \tau(\mu, k),$$

where b_μ^k are given numbers.

Lemma 7.2. *We assume that the condition*

$$\|\tilde{\varepsilon}_p\| < 1 \tag{7.7}$$

is fulfilled for a certain natural number p. Then system (7.6) is uniquely solvable and its solution w_μ^k, $1 \leq \mu \leq M$, $1 \leq k \leq n_\mu$, satisfies the inequality

$$\|w\| \leq \frac{p}{1 - \|\tilde{\varepsilon}_p\|}\|b\|. \tag{7.8}$$

Proof.   We can use Lemma 7.1 to establish the unique solvability of system (7.6) under condition (7.7) in the same way as we established the unique solvability of system (4.8) under condition (4.12) (see the concluding part of the proof of Lemma 4.2 in Sec. 6).

Let us prove inequality (7.8). We set

$$\tilde{z}_{0\mu}^k = w_\mu^k, \qquad b_{0\mu}^k = 0, \tag{7.9}$$

where w_μ^k is a solution of system (7.6), and find the numbers $\tilde{z}_{\nu\mu}^k$ from the recurrent formula (7.5), $1 \leq \mu \leq M$, $1 \leq k \leq n_\mu$, $\nu = 1, 2, \ldots$. Then, according to (7.6), (7.5), we have

$$w_\mu^k = b_{\nu\mu}^k + \tilde{z}_{\nu\mu}^k, \quad \nu = 1, 2, \ldots, \tag{7.10}$$

where

$$b_{\nu\mu}^k = b_\mu^k + \beta_\tau \sum_{m=1}^{n_\tau} b_{\nu-1,\tau}^m \tilde{R}_{\tau\mu}^{km}, \tag{7.11}$$

$$1 \leq \mu \leq M, \quad 1 \leq k \leq n_\mu, \quad \tau = \tau(\mu, k).$$

The inequality $\|b_p\| \leq p\|b\|$ is satisfied by virtue of (7.1), (7.2), (7.9), and (7.11), and, according to Lemma 7.1, we have $\|\tilde{z}_p\| \leq \|\tilde{\varepsilon}_p\| \cdot \|\tilde{z}_0\| = \|\tilde{\varepsilon}_p\| \cdot \|w\|$. This relation and (7.10) imply

$$\|w\| \leq \|b_p\| + \|\tilde{z}_p\| \leq p\|b\| + \|\tilde{\varepsilon}_p\| \cdot \|w\|.$$

Consequently, inequality (7.8) is satisfied under condition (7.7).

Lemma 7.2 is proved.

Remark 7.1. The estimate

$$\|u\| \leq c_0' \frac{M+1}{1 - \|\varepsilon_{M+1}\|},$$

where the constant

$$c_0' = 2 \max_{1 \le \mu \le M} \sup_{(r_\mu, \theta_\mu) \in T_\mu} |Q_\mu(r_\mu, \theta_\mu)| \tag{7.12}$$

does not depend on n, is valid for the solution of system (4.8) (under condition (4.12)) by virtue of (6.6), (6.34) and Lemma 7.2.

We set $w_{0\mu}^k = 0$,

$$w_{\nu\mu}^k = b_\mu^k + \beta_\tau \sum_{m=1}^{n_\tau} w_{\nu-1,\tau}^m \tilde{R}_{\tau\mu}^{km}, \quad \nu = 1, 2, \ldots, \tag{7.13}$$

where $1 \le \mu \le M$, $1 \le k \le n_\mu$, $\tau = \tau(\mu, k)$.

Lemma 7.3. *Under the hypothesis of Lemma 7.2, the successive approximations (7.13) converge to the unique solution of system (7.6), i.e.,*

$$\lim_{\nu \to \infty} \|w_\nu - w\| = 0.$$

Lemma 7.3 can be proved by analogy with Lemma 4.4 (see Sec. 6).

Lemma 7.4. *If $b_\mu^k \ge 0$, $1 \le \mu \le M$, $1 \le k \le n_\mu$, and condition (7.7) is fulfilled for some natural number p, then the unique solution of system (7.6) is nonnegative.*

Proof. By virtue of (7.1) and (7.13) we have $w_{\nu\mu}^k \ge 0$, $1 \le \mu \le M$, $1 \le k \le n_\mu$, $\nu = 1, 2, \ldots$. This and Lemma 7.3 imply the statement of Lemma 7.4

Lemma 7.5. *Let us assume that condition (7.7) is fulfilled for a certain natural number p and w_μ^k, $1 \le \mu \le M$, $1 \le k \le n_\mu$, is a solution of system (7.6) in which $b_\mu^k = b \ge 0$ for all μ and k. Suppose the quantities $\tilde{w}_{0\mu}^k = 0$, $b_{\nu\mu}^k$ are defined,*

$$\tilde{w}_{\nu\mu}^k = b_{\nu\mu}^k + \beta_\tau \sum_{m=1}^{n_\tau} \tilde{w}_{\nu-1,\tau}^m \tilde{R}_{\tau\mu}^{km}, \tag{7.14}$$

$1 \le \mu \le M$, $1 \le k \le n_\mu$, $\tau = \tau(\mu, k)$, with $\|b_\nu\| \le b$, $\nu = 1, 2, \ldots$. Then the inequalities

$$\|\tilde{w}_\nu\| \le \|w\|, \quad \nu = 0, 1, \ldots, \tag{7.15}$$

hold true.

Proof. Note, first of all, that according to Lemma 7.4 the solution of system (7.6) is nonnegative, i.e., $w_\mu^k \geq 0$, $1 \leq \mu \leq M$, $1 \leq k \leq n_\mu$. We assume that the inequalities

$$|\tilde{w}_{\nu\mu}^k| \leq w_\mu^k, \quad 1 \leq \mu \leq M, \quad 1 \leq k \leq n_\mu, \tag{7.16}$$

are satisfied for some fixed ν. Taking into account (7.1), (7.14), (7.16), we find that

$$|\tilde{w}_{\nu+1,\mu}^k| \leq |b_{\nu+1,\mu}^k| + \beta_\tau \sum_{m=1}^{n_\tau} |\tilde{w}_{\nu\tau}^m| |\tilde{R}_{\tau\mu}^{km} \leq b + \beta_\tau \sum_{m=1}^{n_\tau} w_\tau^m \tilde{R}_{\tau\mu}^{km} = w_\mu^k,$$

where $1 \leq \mu \leq M$, $1 \leq k \leq n_\mu$. Bearing in mind that conditions (7.16) are trivially fulfilled for $\nu = 0$, we find, by induction, that inequalities (7.16) are valid for $\nu = 1, 2, \ldots$. Consequently, inequalities (7.15) hold true. We have proved Lemma 7.5.

7.2. Verification of the Criterion of Solvability of an Algebraic Problem

Let us verify criterion (4.12) of the unique solvability of system (4.8) when there are rounding-off errors in computations. We assume that the numbers $\tilde{R}_{\tau\mu}^{km}$, together with conditions (7.1), (7.2), satisfy the inequalities

$$\alpha_\tau \pi |\tilde{R}_{\tau\mu}^{km} - R_{\tau\mu}^{km}| \leq \delta, \tag{7.17}$$

where $R_{\tau\mu}^{km}$ are coefficients (4.7), $\alpha_\tau \pi \leq 2\pi$ is the angular magnitude of the block T_τ (see Remark 3.7), $1 \leq \mu \leq M$, $1 \leq k \leq n_\mu$, $\tau = \tau(\mu, k)$, $1 \leq m \leq n_\tau$, $\delta > 0$ is a given number, i.e., the quantities $\tilde{R}_{\tau\mu}^{km}$ are the approximations of coefficients (4.7) with an accuracy to within $\delta/\alpha_\tau \pi$.

In practice, instead of quantities (4.9), (4.10), we calculate not even quantities (7.3), (7.4), but some numbers

$$e_{0\mu}^k = 1, \tag{7.18}$$

$$e_{\nu\mu}^k = \beta_\tau \sum_{m=1}^{n_\tau} e_{\nu-1,\tau}^m \tilde{R}_{\tau\mu}^{km} + \xi_{\nu\mu}^k, \tag{7.19}$$

where $\xi_{\nu\mu}^k$ are rounding-off errors, $1 \le \mu \le M$, $1 \le k \le n_\mu$, $\tau = \tau(\mu, k)$, $\nu = 1, 2, \dots$. We require that the rounding-off errors satisfy the condition

$$\|\xi_\nu\| \le \delta, \quad \nu = 1, 2, \dots. \tag{7.20}$$

We set

$$\delta_{\nu\mu}^k = e_{\nu\mu}^k - \varepsilon_{\nu\mu}^k, \quad \varkappa_{\nu\mu}^k = \tilde{\varepsilon}_{\nu\mu}^k - \varepsilon_{\nu\mu}^k. \tag{7.21}$$

Then, by virtue of (4.9), (4.10), (7.3), (7.4), (7.18), (7.19), we have $\delta_{0\mu}^k = \varkappa_{0\mu}^k = 0$,

$$\delta_{\nu\mu}^k = \beta_\tau \sum_{m=1}^{n_\tau} \delta_{\nu-1,\tau}^m \tilde{R}_{\tau\mu}^{km} + \beta_\tau \sum_{m=1}^{n_\tau} \varepsilon_{\nu-1,\tau}^m (\tilde{R}_{\tau\mu}^{km} - R_{\tau\mu}^{km}) + \xi_{\nu\mu}^k,$$

$$\varkappa_{\nu\mu}^k = \beta_\tau \sum_{m=1}^{n_\tau} \varkappa_{\nu-1,\tau}^m \tilde{R}_{\tau\mu}^{km} + \beta_\tau \sum_{m=1}^{n_\tau} \varepsilon_{\nu-1,\tau}^m (\tilde{R}_{\tau\mu}^{km} - R_{\tau\mu}^{km}),$$

$1 \le \mu \le M$, $1 \le k \le n_\mu$, $\tau = \tau(\mu, k)$, $\nu = 1, 2, \dots$. On the basis of (4.2), (6.61), (7.1), (7.2), (7.17), (7.20), we arrive, by induction, at inequalities

$$\|\delta_\nu\| \le 2\nu\delta, \quad \|\varkappa_\nu\| \le \nu\delta, \quad \nu = 0, 1, \dots,$$

from which, with due account of (7.21), we find, in turn, that

$$\|\varepsilon_\nu\| \le \|e_\nu\| + 2\nu\delta, \tag{7.22}$$

$$\|e_\nu\| \le \|\varepsilon_\nu\| + 2\nu\delta, \tag{7.23}$$

$$\|\tilde{\varepsilon}_\nu\| \le \|\varepsilon_\nu\| + \nu\delta, \quad \nu = 0, 1, \dots. \tag{7.24}$$

Remark 7.2. By virtue of (7.22), for criterion (4.12) of the unique solvability of system (4.8) to be satisfied, it is sufficient that the inequality

$$\|e_{M+1}\| + 2(M + 1)\delta < 1 \tag{7.25}$$

be valid.

Lemma 7.6. *If the quantity* $\delta > 0$, *entering into inequalities* (7.17), (7.20) *and characterizing the accuracy of computations, satisfies the condition*

$$\delta < \frac{1 - \sigma_0}{4(M + 1)}, \tag{7.26}$$

where $\sigma_0 < 1$ *is the left-hand side of inequality* (4.13), *then conditions* (4.12) *and* (7.25) *are simultaneously fulfilled or not fulfilled (independently of* n).

Lemma 7.6 follows from Lemmas 4.2, 4.3 and inequalities (7.22) and (7.23).

Remark 7.3. Lemma 7.6 testifies to the correctness of criterion (4.12) of the unique solvability of system (4.8).

Remark 7.4. If system (4.8) is uniquely solvable for some fixed n and, hence, $\|\varepsilon_{M+1}\| \leq \sigma_0 < 1$ according to Lemma 4.3, then inequality (7.25) is obviously satisfied even if δ satisfies the inequality

$$\delta < \frac{1 - \|\varepsilon_{M+1}\|}{4(M + 1)},$$

in which, the right-hand side is, as a rule, greater than the right-hand side of inequality (7.26).

Lemma 7.7. *If* δ *satisfies condition* (7.26) *and inequality* (7.25) *holds true for some* n, *then*

$$\|\tilde{\varepsilon}_{M+1}\| < (1 + \sigma_0)/2 \tag{7.27}$$

for a given n.

Proof.   On the basis of Lemmas 7.6, 4.2, 4.3 we have $\|\varepsilon_{M+1}\| \leq \sigma_0$. This inequality and (7.24), (7.26) imply (7.27).

7.3.   The Stability of the Block Method in a Uniform Metric

Suppose that we have to find an approximate solution of the boundary-value problem (1.1), (1.2) with a uniform accuracy of $\varepsilon > 0$. The problem is complicated by the fact that, in practice, computations are not carried out exactly. Assume that instead of (4.15) we calculate

$$\tilde{u}^k_{\nu\mu} = \tilde{a}^k_\mu + \beta_\tau \sum_{m=1}^{n_\tau} \tilde{u}^m_{\nu-1,\tau} \tilde{R}^{km}_{\tau\mu} + \eta^k_{\nu\mu}, \tag{7.28}$$

where $\tilde{u}_{0\mu}^k = 0$, $\tilde{R}_{\tau\mu}^{km}$ are approximate values of coefficients (4.7) satisfying conditions (7.1), (7.2), (7.17), $\tilde{a}_\mu^k$ are approximate values of the free terms (4.14), $\eta_{\nu\mu}^k$ are rounding-off errors, with

$$\|\tilde{a} - a\| \le \delta, \quad \|\eta_\nu\| \le \delta, \tag{7.29}$$

$\delta > 0$ is the same as in inequalities (7.17), (7.20), $1 \le \mu \le M$, $1 \le k \le n_\mu$, $\tau = \tau(\mu, k)$ is defined by formula (4.4), $\nu = 1, 2, \ldots$.

Proceeding from a given $\varepsilon > 0$, we must indicate n, $\delta > 0$, and the number of iterations $\nu = \nu^*$ under which the approximate solution

$$\tilde{u}_{\mu n}(r_\mu, \theta_\mu) = Q_\mu(r_\mu, \theta_\mu) + \beta_\mu \sum_{k=1}^{n_\mu} (\tilde{u}_{\nu^* \cdot \mu}^k - Q_\mu^k) R_\mu(r_\mu, \theta_\mu, \theta_\mu^k) \tag{7.30}$$

constructed by formula (5.1) in which the quantities $\tilde{u}_{\nu^* \cdot \mu}^k$, $1 \le \mu \le M$, $1 \le k \le n_\mu$, have been substituted for the exact solution of system (4.8), satisfies the inequality

$$\max_{(r_\mu, \theta_\mu) \in \overline{T}_\mu^{**}} |\tilde{u}_{\mu n}(r_\mu, \theta_\mu) - u(r_\mu, \theta_\mu)| < \varepsilon, \tag{7.31}$$

where $T_\mu^{**} \supset T_\mu^*$ is the intermediate block (5.2) and u is a solution of the boundary-value problem (1.1), (1.2), $1 \le \mu \le M$.

Let us estimate the orders of the required quantities n, δ, and ν^* with respect to ε ($\varepsilon \to +0$). We first choose n for which system (4.8) is uniquely solvable and the condition

$$c_0^* \exp\{-d_0 n\} < \varepsilon/2 \tag{7.32}$$

is satisfied, in which c_0^*, $d_0 > 0$ are constants from (6.57). Clearly,

$$n = O(|\ln \varepsilon|) \tag{7.33}$$

according to Lemma 4.1 (Theorem 6.1) and (7.32).

As concerns δ, we assume, for the time being, that it satisfies condition (7.26) which guarantees, according to Lemma 7.6, the fulfillment of inequality (7.25) since system (4.8) is uniquely solvable for the chosen n.

We set $\Delta_{\nu\mu}^k = \tilde{u}_{\nu\mu}^k - u_{\nu\mu}^k$. Then, according to (4.15), (7.28), we have $\Delta_{0\mu}^k = 0$ and

$$\Delta_{\nu\mu}^k = \beta_\tau \sum_{m=1}^{n_\tau} \Delta_{\nu-1,\tau}^m \tilde{R}_{\tau\mu}^{km} + (\tilde{a}_\mu^k - a_\mu^k)$$

$$+ \beta_\tau \sum_{m=1}^{n_\tau} u_{\nu-1,\tau}^m (\tilde{R}_{\tau\mu}^{km} - R_{\tau\mu}^{km}) + \eta_{\nu\mu}^k, \qquad (7.34)$$

where $1 \leq \mu \leq M$, $1 \leq k \leq n_\mu$, $\tau = \tau(\mu, k)$ and $\nu = 1, 2, \ldots$.

We require that the inequalities

$$\|u_{\nu^*} - u\| \leq \varepsilon/4\tilde{c}, \qquad (7.35)$$

$$\|\Delta_{\nu^*}\| \leq \varepsilon/4\tilde{c}, \qquad (7.36)$$

where u is a solution of system (4.8), be satisfied, as well as

$$\tilde{c} = \pi \max_{1 \leq \mu \leq M} \alpha_\mu \max_{(r_\mu, \theta_\mu) \in \overline{T}_\mu^{**}} \max_{|\eta| < \infty} |R_\mu(r_\mu, \theta_\mu, \eta)|, \qquad (7.37)$$

and, according to (3.12)–(3.18), (5.2), $0 < \tilde{c} < \infty$.

By Lemmas 4.3 and 4.4, inequality (7.35) is satisfied for some

$$\nu^* = O(|\ln \varepsilon|) \qquad (7.38)$$

independent of n (for which system (4.8) is uniquely solvable). Thus n and ν^* have been chosen.

Let us find δ for which requirement (7.36) is fulfilled. We apply Lemma 7.5 to quantities (4.15) which play the part of quantities (7.14). We set $p = M + 1$, $b = c_0'$, where c_0' is expressed by (7.12), substitute coefficients (4.7) for coefficients $\tilde{R}_{\tau\mu}^{km}$ in (7.6) and, correspondingly, replace condition (7.7) by condition (4.12) which is fulfilled for the chosen n. Then, using also Lemmas 7.2 and 4.3, we get

$$\|u_\nu\| \leq \frac{M+1}{1 - \|\varepsilon_{M+1}\|} c_0' \leq \frac{M+1}{1-\sigma_0} c_0',$$

where $\nu = 0, 1, \ldots$. This inequality and (4.2), (7.17), (7.29) yield an inequality

$$|(\tilde{a}_\mu^k - a_\mu^k) + \beta_\tau \sum_{m=1}^{n_\tau} u_{\nu\tau}^m (\tilde{R}_{\tau\mu}^{km} - R_{\tau\mu}^{km}) + \eta_{\nu\mu}^k| \leq$$

$$\leq \delta\left(2 + \frac{M+1}{1-\sigma_0}c_0'\right), \tag{7.39}$$

$$\nu = 0, 1, \ldots, \qquad 1 \leq \mu \leq M, \qquad 1 \leq k \leq n_\mu, \qquad \tau = \tau(\mu, k).$$

Since we have imposed condition (7.26) on δ, inequality (7.25) is satisfied for chosen n (for which system (4.8) is uniquely solvable) by virtue of Lemmas 4.2 and 7.6. Consequently, on the basis of (7.34), (7.39) and Lemmas 7.5, 7.2, 7.7 (in Lemma 7.5 we set $p = M+1$ and take the right-hand side of (7.39) as b), we have

$$\|\Delta_\nu\| \leq \|w\| \leq \frac{M+1}{1-\|\tilde{\mathscr{E}}_{M+1}\|}\delta\left(2 + \frac{M+1}{1-\sigma_0}c_0'\right) \leq c_0''\delta, \qquad \nu = 0, 1, \ldots,$$

where the constant

$$c_0'' = 2\frac{M+1}{1-\sigma_0}\left(2 + \frac{M+1}{1-\sigma_0}c_0'\right)$$

does not depend on ν and n.

We assume, for simplicity, that the given ε satisfies the inequalities $0 < \varepsilon \leq 1$. Then, evidently, we can set

$$\delta = \frac{\varepsilon}{(4\tilde{c}+1)c_0''}, \tag{7.40}$$

and thus simultaneously satisfy (7.36) and (7.26). Thus the chosen δ has an exact order ε.

Thus, in accordance with (4.2), (5.1), (7.35)–(7.37) we have

$$\max_{(r_\mu, \theta_\mu) \in \overline{T}_\mu^{\bullet\bullet}} |\tilde{u}_{\mu n}(r_\mu, \theta_\mu) - u_{\mu n}(r_\mu, \theta_\mu)|$$

$$\leq \beta_\mu \max_{(r_\mu, \theta_\mu) \in \overline{T}_\mu^{\bullet\bullet}} \left|\sum_{k=1}^{n_\mu}(\tilde{u}_{\nu\bullet\mu}^k - u_\mu^k)R_\mu(r_\mu, \theta_\mu, \theta_\mu^k)\right| \leq \tilde{c}\sum_{k=1}^{n_\mu}\frac{|\tilde{u}_{\nu\bullet\mu}^k - u_\mu^k|}{n_\mu}$$

$$\leq \tilde{c}\left(\sum_{k=1}^{n_\mu}\frac{|\Delta_{\nu\bullet\mu}^k|}{n_\mu} + \sum_{k=1}^{n_\mu}\frac{|u_{\nu\bullet\mu}^k - u_\mu^k|}{n_\mu}\right) \leq \frac{\varepsilon}{2},$$

where $1 \leq \mu \leq M$. This relation and (6.57), (7.32) yield inequality (7.31) for the chosen n, δ and $\nu = \nu^*$.

Remark 7.5. The existence of inequality (7.31) under requirements (7.17), (7.29) imposed upon the accuracy of calculations, where δ has an exact order ε in accordance with (7.40), means that the block method is stable in a uniform metric. By virtue of (6.6) and (6.34), the additional requirements (7.1), (7.2) imposed on the approximate values $\tilde{R}^{km}_{\tau\mu}$ of coefficients (4.7) can be easily satisfied in practical computations.

Remark 7.6. For n, δ, and ν^* chosen from $\varepsilon \in (0,1]$ in the way described above, on the basis of inequality (7.31) and Lemmas 3.4 and 6.12 we have inequalities of form (5.3)–(5.5), in which the function $u_{\mu n}(r_\mu, \theta_\mu)$ on the left-hand side is replaced by function (7.30) and $\exp\{-d_0 n\}$ on the right-hand side is replaced by ε, and the constant c_p on the right-hand side does not depend on ε (see the concluding part of the proof of Theorem 5.1 in Sec. 6).

Remark 7.7. When the sum in (7.28) is calculated with the aid of a computer with a fixed number length, the rounding-off error increases with the growth of n, but no faster than in proportion to n. Therefore, in practical calculations on a computer, the error in the result first rapidly decreases according to an exponential law with the growth of n and then, when the errors inherent in the block method and rounding-off errors level off, the accuracy of the result slowly decreases (see the case $t = 4$ in Table 24.1).

7.4.   The Labor of Computations Required by the Block Method

Let us estimate the number of arithmetic operations necessary for finding the approximate solution (7.30) which satisfies inequality (7.31), $1 \le \mu \le M$. According to (7.33), (7.38), and the inequality $n_\mu \le 4n$ (see (4.1)) the number N_1 of arithmetic operations involved in calculating $\tilde{u}^k_{\nu \bullet \mu}$, $1 \le \mu \le M$, $1 \le k \le n_\mu$, by formula (7.28) is

$$N_1 = O(|\ln^3 \varepsilon|).$$

In addition, when using formulas (4.6), (4.7), (4.14) to find the coefficients $\tilde{R}^{km}_{\tau\mu}$, $\tilde{a}^k_\mu$, which are $O(\ln^2 \varepsilon)$ in number, we can restrict our calculations to $O(|\ln \varepsilon|)$ different values of elementary functions with an accuracy of $O(\delta)$. For our purpose we take the kernel of Poisson's integral, in terms of which kernels (3.15)–(3.17) are expressed, in the form

$$R(r, \theta, \eta) = \frac{1 - r^2}{2\pi(1 - 2r(\cos\theta\cos\eta + \sin\theta\sin\eta) + r^2)} \qquad (7.41)$$

and repeatedly use the values of $\cos\eta$ and $\sin\eta$ calculated by formula (4.7) for different θ.

By means of $O(1)$ operations the calculation of elementary functions are reduced to the summation of power series with coefficients that are sufficiently simply constructed. These series converge in a geometric progression with a common ratio bounded by a number smaller than unity. Therefore, according to (7.40), every value of an elementary function can be found by means of $O(|\ln\delta|) = O(|\ln\varepsilon|)$ arithmetic operations carried out with an accuracy of $O(\varepsilon|\ln\varepsilon|^{-1})$. Thus the total number of arithmetic operations needed to calculate the coefficients $\bar{R}_{\tau\mu}^{km}$, $\tilde{a}_\mu^k$ is

$$N_2 = O(\ln^2\varepsilon).$$

Thus we need
$$N_1 + N_2 = O(|\ln^3\varepsilon|) \qquad (7.42)$$

arithmetic operations.

Remark 7.8. When calculating the coefficients $\bar{R}_{\tau\mu}^{km}$ by formula (7.28), we need not retain them in the computer storage but calculate them in iterations for every pair μ, k anew with the use of the kernel of Poisson's integral in form (7.41). In order to find the quantities $\tilde{u}_{\nu\bullet\mu}^k$, where $1 \leq \mu \leq M$, $1 \leq k \leq n_\mu$, we shall need only $O(|\ln\varepsilon|)$ storage locations and the order of the total number of arithmetic operations needed will not change relative to ε.

Remark 7.9. To make the use of the approximate solution (7.30) more convenient, it is expedient to retain in the computer storage not only the quantities $\tilde{u}_{\nu\bullet\mu}^k$, $1 \leq \mu \leq M$, $1 \leq k \leq n_\mu$, but also the corresponding tables for $\cos\eta$ and $\sin\eta$ with the step $\pi/2n_\mu$ if $1 \leq \mu \leq N$ and $\nu_{\mu-1} + \nu_\mu = 1$, or with the step π/n_μ in other cases. The values Q_τ^m, $1 \leq m \leq n_\tau$, are also stored (see (4.6)). The total number of $O(|\ln\varepsilon|)$ values is retained in the computer storage. One value of the approximate solution can be found from formula (7.30) (i.e., from formula (5.1), where u_μ^k are replaced by $\tilde{u}_{\nu\bullet\mu}^k$) at an arbitrary point of the closed basic block $\overline{T}_\mu^*$ with an error of $O(\varepsilon)$ by means of $O(|\ln\varepsilon|)$ arithmetic operations.

8. Approximation of a Conjugate Harmonic Function on Blocks

In this paragraph we consider the approximate determination of a harmonic function conjugate to the solution of the boundary-value problem (1.1), (1.2) only locally, namely, on the basic blocks T_μ^*, $1 \le \mu \le M$, which are part of the covering of the polygon Ω. If the polygon Ω is multiply-connected, the conjugate function will be many-valued on Ω in the general case. But since the blocks (sectors of disks, half-disks, and disks) are simply-connected subdomains, the conjugate harmonic function is defined, separately on each block, as a definite function which contains only an arbitrary additive constant. The approximations of a conjugate harmonic function obtained on blocks in the form of elementary harmonic functions will be effectively used in the next paragraphs for an approximate solution of some problems, in particular, Neumann's problem (see also Remark 8.3).

As an approximation of the function conjugate to the solution u of the boundary-value problem (1.1), (1.2) on the basic block T_μ^*, $1 \le \mu \le M$, it is natural to take the harmonic function conjugate to the approximate solution (5.1) of problem (1.1), (1.2) on this block which can be written down explicity. We first seek an expression for a harmonic function conjugate to (5.1) for all possible cases and then investigate the problem of the accuracy of the approximate conjugate harmonic function obtained on blocks.

We denote by $S_\mu(r_\mu, \theta_\mu)$ the harmonic conjugate of the carrier function $Q_\mu(r_\mu, \theta_\mu)$ appearing as a term in (5.1). We understand S_μ to be a certain, quite definite, function which does not contain an arbitrary additive constant.

In accordance with (3.2)–(3.9), depending on the values of the parameters ν_{j-1} and ν_j, the function S_j, $1 \le j \le N$, is defined on the closed basic block-sector $\overline{T}_j^*$ as follows:

(1) for $\nu_{j-1} = \nu_j = 1$ in the form

$$S_j(r_j, \theta_j) = \frac{b_{j0} - a_{j0}}{\alpha_j \pi} \ln r_j$$

$$+ \sum_{k=1}^{m_{j-1}} a_{jk} \bar{\xi}_{jk}(r_j, \theta_j) - \sum_{k=1}^{m_j} b_{jk} \bar{\xi}_{jk}(r_j, \alpha_j \pi - \theta_j), \quad (8.1)$$

where

$$\bar{\xi}_{jk}(r_j, \theta_j) = \begin{cases} r_j^k \dfrac{\theta_j \sin k\theta_j - \ln r_j \cos k\theta_j}{\alpha_j \pi \cos k\alpha_j \pi}, & \sin k\alpha_j \pi = 0, \\[3mm] -r_j^k \dfrac{\cos k\theta_j}{\sin k\alpha_j \pi}, & |\sin k\alpha_j \pi| \geq \dfrac{1}{2}, \\[3mm] \dfrac{\sigma_{jk} r_j^{k'} \cos k'\theta_j - r_j^k \cos k\theta_j}{\sin k\alpha_j \pi}, & 0 < |\sin k\alpha_j \pi| < \dfrac{1}{2}, \end{cases} \quad (8.2)$$

$k' = [k\alpha_j + 1/2]/\alpha_j$, $[\ ]$ is the sign of the integer part, $\sigma_{jk} = (6r_{j0}/5)^{k-k'}$, $\alpha_j \pi$ is the magnitude of the vertex angle of the sector T_j^*,

(2) for $\nu_{j-1} = \nu_j = 0$ in the form

$$S_j(r_j, \theta_j) = \sum_{k=0}^{m_{j-1}} a_{jk} \bar{\eta}_{jk}(r_j, \theta_j) - \sum_{k=0}^{m_j} b_{jk} \bar{\eta}_{jk}(r_j, \alpha_j \pi - \theta_j), \quad (8.3)$$

where

$$\bar{\eta}_{jk}(r_j, \theta_j) =$$

$$= \begin{cases} r_j^{k+1} \dfrac{\ln r_j \sin(k+1)\theta_j + \theta_j \cos(k+1)\theta_j}{(k+1)\alpha_j \pi \cos(k+1)\alpha_j \pi}, & \sin(k+1)\alpha_j \pi = 0, \\[3mm] r_j^{k+1} \dfrac{\sin(k+1)\theta_j}{(k+1)\sin(k+1)\alpha_j \pi}, & |\sin(k+1)\alpha_j \pi| \geq \dfrac{1}{2}, \\[3mm] \dfrac{r_j^{k+1} \sin(k+1)\theta_j - \sigma_{jk} r_j^{k'} \sin k'\theta_j}{(k+1)\sin(k+1)\alpha_j \pi}, & 0 < |\sin(k+1)\alpha_j \pi| < \dfrac{1}{2}, \end{cases} \quad (8.4)$$

$$k' = [(k+1)\alpha_j + 1/2]/\alpha_j, \qquad \sigma_{jk} = (6r_{j0}/5)^{k+1-k'},$$

(3) for $\nu_{j-1} = 0$, $\nu_j = 1$ in the form

$$S_j(r_j, \theta_j) = \sum_{k=0}^{m_{j-1}} a_{jk} \bar{\vartheta}_{jk}(r_j, \theta_j) - \sum_{k=1}^{m_j} b_{jk} \bar{\varkappa}_{jk}(r_j, \alpha_j \pi - \theta_j), \quad (8.5)$$

where

$$\bar{\vartheta}_{jk}(r_j, \theta_j) =$$

$$= \begin{cases} r_j^{k+1} \dfrac{\theta_j \sin(k+1)\theta_j - \ln r_j \cos(k+1)\theta_j}{(k+1)\alpha_j\pi \sin(k+1)\alpha_j\pi}, & \cos(k+1)\alpha_j\pi = 0, \\[3mm] r_j^{k+1} \dfrac{\cos(k+1)\theta_j}{(k+1)\cos(k+1)\alpha_j\pi}, & |\cos(k+1)\alpha_j\pi| \geq \dfrac{1}{2}, \\[3mm] \dfrac{r_j^{k+1}\cos(k+1)\theta_j - \sigma_{jk}r_j^{k'}\cos k'\theta_j}{(k+1)\cos(k+1)\alpha_j\pi}, & 0 < |\cos(k+1)\alpha_j\pi| < \dfrac{1}{2}, \end{cases} \quad (8.6)$$

$$k' = (2[(k+1)\alpha_j] + 1)/2\alpha_j, \qquad \sigma_{jk} = (6r_{j0}/5)^{k+1-k'},$$

$$\overline{\varkappa}_{jk}(r_j, \theta_j) =$$

$$= \begin{cases} -r_j^k \dfrac{\ln r_j \sin k\theta_j + \theta_j \cos k\theta_j}{\alpha_j\pi \sin k\alpha_j\pi}, & \cos k\alpha_j\pi = 0, \\[3mm] r_j^k \dfrac{\sin k\theta_j}{\cos k\alpha_j\pi}, & |\cos k\alpha_j\pi| \geq \dfrac{1}{2}, \\[3mm] \dfrac{r_j^k \sin k\theta_j - \tau_{jk}r_j^{k''}\sin k''\theta_j}{\cos k\alpha_j\pi}, & 0 < |\cos k\alpha_j\pi| < \dfrac{1}{2}, \end{cases} \quad (8.7)$$

$$k'' = (2[k\alpha_j] + 1)/2\alpha_j, \qquad \tau_{jk} = (6r_{j0}/5)^{k-k''},$$

(4) for $\nu_{j-1} = 1$, $\nu_j = 0$ in the form

$$S_j(r_j, \theta_j) = \sum_{k=1}^{m_{j-1}} a_{jk}\overline{\varkappa}_{jk}(r_j, \theta_j) - \sum_{k=0}^{m_j} b_{jk}\bar{\vartheta}_{jk}(r_j, \alpha_j\pi - \theta_j). \quad (8.8)$$

According to (3.10), for $N < q \leq L$ we have

$$S_q(r_q, \theta_q) = \sum_{k=0}^{m_{j(q)}} c_{qk}r_q^k \left(\nu_{j(q)}\sin k\theta_q - \bar{\vartheta}_{j(q)}r_q\frac{\cos(k+1)\theta_q}{k+1}\right). \quad (8.9)$$

Finally, in accordance with (3.11), we set

$$S_p(r_p, \theta_p) = 0 \quad (8.10)$$

for $L < p \leq M$.

To find the conjugate of the kernel R_μ which is under the sign of the sum in formula (5.1), we do the following.

We set

$$I(m, m, r, \theta, \eta) = I(r, \theta, \eta) + (-1)^m I(r, \theta, -\eta), \qquad (8.11)$$

$$\begin{aligned} I(1 - m, m, r, \theta, \eta) &= I(m, m, r, \theta, \eta) \\ &\quad - (-1)^m I(m, m, r, \theta, \pi - \eta), \end{aligned} \qquad (8.12)$$

where $m = 0, 1$, and

$$I(r, \theta, \eta) = \frac{r \sin(\theta - \eta)}{\pi(1 - 2r \cos(\theta - \eta) + r^2)} \qquad (8.13)$$

is the conjugate of kernel (3.14) of Poisson's integral for a unit disk (see [23]).

Let us now define the following kernels in accordance with the type of the block:

$$I_p(r_p, \theta_p, \eta) = I(r_p/r_{p0}, \theta_p, \eta), \quad L < p \le M, \qquad (8.14)$$

$$I_q(r_q, \theta_q, \eta) = I(\nu_{j(q)}, \nu_{j(q)}, \frac{r_q}{r_{q0}}, \theta_q, \eta), \quad N < q \le L, \qquad (8.15)$$

$$I_j(r_j, \theta_j, \eta) = \lambda_j I\left(\nu_{j-1}, \nu_j, \left(\frac{r_j}{r_{j0}}\right)^{\lambda_j}, \lambda_j \theta_j, \lambda_j \eta\right), \quad 1 \le j \le N, \qquad (8.16)$$

where $j(q)$ is the index of the side which the half-disk T_q adjoins, ν_{j-1}, ν_j, $\nu_{j(q)}$ are parameters entering into boundary conditions (1.2), λ_j is quantity (3.18). Kernels (8.14)–(8.16) are the conjugates of kernels (3.15)–(3.17).

On the basis of what has been said we arrive at the following lemma.

Lemma 8.1. Let u_μ^k, $1 \le \mu \le M$, $1 \le k \le n_\mu$, be a solution of the system of linear algebraic equations (4.8).

The function

$$v_{\mu n}(r_\mu, \theta_\mu) = C_\mu + S_\mu(r_\mu, \theta_\mu) + \beta_\mu \sum_{k=1}^{n_\mu}(u_\mu^k - Q_\mu^k)I_\mu(r_\mu, \theta_\mu, \theta_\mu^k), \qquad (8.17)$$

where C_μ is an arbitrary constant and the kernel I_μ, the function S_μ, and, the quantities n_μ, β_μ, θ_μ^k, Q_μ^k are defined by formulas (8.14)–(8.16), (8.1)–(8.10) and (4.1)–(4.3), (4.6) respectively, is a conjugate harmonic function to the approximate solution (5.1) of the boundary-value problem (1.1), (1.2) on the basic block T_μ^*, $1 \le \mu \le M$. This function is continuous up to the boundary of the basic block, except for the vertex P_μ (when $1 \le \mu \le N$) of the sector T_μ^* if the boundary values of the required solution, discontinuous at the vertex, are defined on its sides.

We can also verify the validity of Lemma 8.1 if we use the Cauchy-Riemann conditions which have the form

$$\frac{\partial u}{\partial r} = \frac{1}{r}\frac{\partial v}{\partial \theta}, \qquad \frac{1}{r}\frac{\partial u}{\partial \theta} = -\frac{\partial v}{\partial r}.$$

in the polar coordinate system r, θ.

Denoting by v_μ a branch, continuous and single-valued on the extended block T_μ, of the harmonic function conjugate to the solution u of the boundary-value problem (1.1), (1.2). In accordance with (3.31) (Theorem 3.1) this function can be represented on T_μ in the form

$$v_\mu(r_\mu, \theta_\mu) = C_\mu' + S_\mu(r_\mu, \theta_\mu)$$
$$+ \int_0^{\alpha_\mu \pi} (u(r_{\mu 0}, \eta) - Q_\mu(r_{\mu 0}, \eta)) I_\mu(r_\mu, \theta_\mu, \eta) \, d\eta, \quad (8.18)$$

where C_μ' is a constant, S_μ is a function conjugate to Q_μ that can be expressed by formulas (8.1)–(8.10), I_μ is a kernel defined by formulas (8.14)–(8.16) which is conjugate to the kernel R_μ.

Remark 8.1. It is easy to see that function (8.18), as well as function (8.17), is continuous up to the boundary of the basic block $T_\mu^* \subset T_\mu$, except the only case when T_μ^* is a sector $(1 \le \mu \le N)$ with $\nu_{\mu-1} = \nu_\mu = 1$ and the boundary values at the vertex P_μ defined in conditions (1.2), are discontinuous. In this case the continuity of the function S_μ, determined from (8.1), where $j = \mu$, at vertex P_μ is violated at the expense of the term $(b_{\mu 0} - a_{\mu 0}) \ln r_\mu / \alpha_\mu \pi$ with the coefficient $b_{\mu 0} - a_{\mu 0} \ne 0$.

Lemma 8.2. *Suppose that v_μ is a branch, continuous and single-valued on the basic block T_μ^*, of the harmonic function conjugate to the*

solution u of the boundary-value problem (1.1), (1.2), $v_{\mu n}$ is function
(8.17) where $1 \leq \mu \leq M$. Then the following three statements hold
true.

(1) The inequality

$$\left| \frac{\partial^p}{\partial x^{p-q} \partial y^q} (v_{\mu n}(r_\mu, \theta_\mu) - v_\mu(r_\mu, \theta_\mu)) \right| \leq c_p \exp\{-d_0 n\} \qquad (8.19)$$

holds true on $\overline{T}_\mu^*$, first, when $N < \mu \leq M$, for $p \geq 1$, second, when
$1 \leq \mu \leq N$, λ_μ is integer and $\nu_{\mu-1}$ and ν_μ are arbitrary for $p \geq \lambda_\mu$.

(2) The estimate

$$\left| \frac{\partial^p}{\partial x^{p-q} \partial y^q} (v_{\mu n}(r_\mu, \theta_\mu) - v_\mu(r_\mu, \theta_\mu)) \right| \leq c_p r_\mu^{\lambda_\mu - p} \exp\{-d_0 n\} \qquad (8.20)$$

holds true on $\overline{T}_\mu^*$ for $1 \leq \mu \leq N$ and any $\nu_{\mu-1}$, ν_μ, and λ_μ, if $1 \leq p < \lambda_\mu$.

(3) The inequality

$$\left| \frac{\partial^p}{\partial x^{p-q} \partial y^q} (v_{\mu n}(r_\mu, \theta_\mu) - v_\mu(r_\mu, \theta_\mu)) \right| \leq c_p r_\mu^{\lambda_\mu - p} \exp\{-d_0 n\} \qquad (8.21)$$

holds true on $\overline{T}_\mu^* \setminus P_\mu$ for $1 \leq \mu \leq N$, noninteger λ_μ and any $\nu_{\mu-1}$
and ν_μ when $p > \lambda_\mu$.

Everywhere $0 \leq q \leq p$, λ_μ is quantity (3.18), c_p, $p = 1, 2, \ldots$, and
$d_0 > 0$ are constants independent of $r_\mu = r_\mu(x, y)$, $\theta_\mu = \theta_\mu(x, y)$ and
of n for which system (4.8) has a unique solution, P_μ in (3) is the
vertex of the block-sector T_μ^*.

Lemma 8.2 follows from Theorem 5.1 and the Cauchy-Riemann
conditions since the functions $(u_{\mu n} - u)$ and $(v_{\mu n} - v_\mu)$ are conjugate
harmonic functions on T_μ^*, $1 \leq \mu \leq M$.

Remark 8.2. For $1 \leq \mu \leq N$ each of the functions $v_{\mu n}$ and v_μ
taken separately, can be nondifferentiable and even not defined at the
vertex P_μ of the sector T_μ^*. However, the definition of the derivatives
of their difference, of the order of $p < \lambda_\mu$, can be completed, by
continuity, at P_μ by a zero, which fact is reflected in inequality (8.20)
(just as in (5.4)).

We set $\omega_\mu^* = P_\mu$ if $1 \leq \mu \leq N$ and functions (8.17), (8.18) are
discontinuous at the vertex P_μ of the block-sector T_μ^*, i.e., $b_{\mu 0} \neq a_{\mu 0}$
(see Remark 8.1). In the other cases we set $\omega_\mu^* = \varnothing$.

Lemma 8.3. *The following inequality holds true:*

$$|(v_{\mu n}(r''_\mu, \theta''_\mu) - v_{\mu n}(r'_\mu, \theta'_\mu)) - (v_\mu(r''_\mu, \theta''_\mu) - v_\mu(r'_\mu, \theta'_\mu))|$$

$$\leq c \exp\{-d_0 n\}, \tag{8.22}$$

where $1 \leq \mu \leq M$, $v_{\mu n}$, v_μ *are functions* (8.17), (8.18), (r'_μ, θ'_μ), $(r''_\mu, \theta''_\mu) \in \overline{T}^*_\mu \setminus \omega^*_\mu$ *are arbitrary points,* c, $d_0 > 0$ *are constants independent of* n *for which system* (4.8) *has a unique solution.*

Proof.   Without violating generality we can assume that $r''_\mu > 0$. We have

$$(v_{\mu n}(r''_\mu, \theta''_\mu) - v_{\mu n}(r'_\mu, \theta'_\mu)) - (v_\mu(r''_\mu, \theta''_\mu) - v_\mu(r'_\mu, \theta'_\mu))$$

$$= (v_{\mu n}(r''_\mu, \theta''_\mu) - v_\mu(r''_\mu, \theta''_\mu)) - (v_{\mu n}(r''_\mu, \theta'_\mu) - v_\mu(r''_\mu, \theta'_\mu))$$

$$+ (v_{\mu n}(r''_\mu, \theta'_\mu) - v_\mu(r''_\mu, \theta'_\mu)) - (v_{\mu n}(r'_\mu, \theta'_\mu) - v_\mu(r'_\mu, \theta'_\mu))$$

$$= \int_{\theta'_\mu}^{\theta''_\mu} \frac{\partial}{\partial \theta_\mu}(v_{\mu n}(r''_\mu, \theta_\mu) - v_\mu(r''_\mu, \theta_\mu)) \, d\theta_\mu$$

$$+ \int_{r'_\mu}^{r''_\mu} \frac{\partial}{\partial r_\mu}(v_{\mu n}(r_\mu, \theta'_\mu) - v_\mu(r_\mu, \theta'_\mu)) \, dr_\mu. \tag{8.23}$$

Since $0 \leq r'_\mu \leq r^*_\mu$, $0 < r''_\mu \leq r^*_\mu$, $|\theta''_\mu - \theta'_\mu| \leq \alpha_\mu \pi$ and the obvious inequalities

$$\left|\frac{\partial}{\partial r_\mu}(v_{\mu n} - v_\mu)\right| \leq \left|\frac{\partial}{\partial x}(v_{\mu n} - v_\mu)\right| + \left|\frac{\partial}{\partial y}(v_{\mu n} - v_\mu)\right|,$$

$$\left|\frac{\partial}{\partial \theta_\mu}(v_{\mu n} - v_\mu)\right| \leq r_\mu\left(\left|\frac{\partial}{\partial x}(v_{\mu n} - v_\mu)\right| + \left|\frac{\partial}{\partial y}(v_{\mu n} - v_\mu)\right|\right)$$

hold true on $\overline{T}^*_\mu \setminus P_\mu$, these relations and (8.23) and Lemma 8.2 (inequalities (8.19)–(8.21) for $p = 1$) easily yield the statements of Lemma 8.3.

Remark 8.3. Evidently, the flux of $\mathrm{grad}\, u$ (u is the solution of the boundary-value problem (1.1), (1.2)) across a piecewise-smooth curve, which may contain a part of the boundary of the polygon Ω, is equal to the increment of the conjugate harmonic function v

on this curve (with the opposite sign). Therefore, in practice it is not advisable to calculate directly the flux of grad$u$ expressed by the corresponding integral along a curve since it is easier to find the increment of the conjugate function. If a curve lies on several basic blocks of the covering of the polygon Ω (see Sec. 2), then we divide the curve into several segments each of which entirely lies on a separate block. On each segment of the curve the increment of the function v is equal to the difference of its values at its endpoints. According to Lemma 8.3, the error of calculation of the increment of the function v within one block $\overline{T}_\mu^*$, $1 \le \mu \le M$, is equal to $O(\exp\{-d_0 n\})$ at the expense of its approximation by the function $v_{\mu n}$. It is assumed that the piecewise-smooth curve indicated above does not contain the vertices of the polygon Ω at which the solution u of problem (1.1), (1.2) is discontinuous.

9.   Neumann's Problem

9.1.   The Solvability of Neumann's Problem on a Polygon

Let us consider, on a simply-connected polygon Ω, Neumann's problem

$$\Delta v = 0 \quad \text{on} \quad \Omega,$$
$$v_n' = \psi_j(s) \quad \text{on} \quad \gamma_j, \qquad j = 1, 2, \ldots, N, \qquad (9.1)$$

where v_n' is the derivative along the inner normal to the side γ_j, $\psi_j(s)$ is a defined algebraic polynomial of the arc length s reckoned along γ_j in the direction of the positive traverse of the boundary of the polygon Ω. We do not impose any local conditions of consistency on the polynomials $\psi_{j-1}(s)$ and $\psi_j(s)$ at every vertex P_j, including the case of an angle of magnitude π or 2π. We assume that the general condition

$$\sum_{j=1}^{N} \int_{\gamma_j} \psi_j(s)\, ds = 0 \qquad (9.2)$$

is fulfilled.

Definition 9.1. *The solution of Neumann's problem (9.1) is a function v with the following properties:*

(1) v satisfies Laplace's equation on Ω,

(2) *v is continuous on the polygon Ω up to its boundary (its vertices inclusive),*

(3) *on the polygon Ω the function v has partial derivatives $\partial v/\partial x$ and $\partial v/\partial y$ continuous up to its boundary, except for some of its vertices,*

(4) *on every side γ_j of the polygon Ω, except, maybe, one or two of its endpoints, the derivative of the function v along the inner normal to γ_j assumes values equal to the values of the given polynomial $\psi_j(s)$.*

This definition of the solution of Neumann's problem conforms with Definition 1.5 of the solution of the mixed boundary-value problem (1.1), (1.2) (see also Definition 1.4 and the explanation to the formulation of problem (1.1), (1.2) given in Sec. 1).

Let us investigate the solvability of the formulated Neumann's problem, which is beyond the scope of the classical case, when the domain has a smooth boundary.

We assume that the covering of the polygon Ω (on which problem (9.1) is stated) by sectors, half-disks, and disks is defined in accordance with requirements I–IV from Sec. 2. This covering will later be used for constructing an approximate solution of problem (9.1) and for proving some assertions.

Lemma 9.1. *Condition (9.2) is necessary for Neumann's problem (9.1) to be solvable.*

Lemma 9.1 can be proved by analogy with Lemma 3.3.

Lemma 9.2. *Suppose that D is a domain, γ is its boundary, $P \in \gamma$ is a point at which there is a tangent and, therefore, a normal to γ. Suppose, furthermore, that there is a closed disk $\overline{K}$ (of a positive radius) which lies entirely in D, except for a single (boundary) point coincident with P. Then, if the function u is harmonic on D, continuous up to the point P, has a derivative u'_n along the inner normal to γ (directed toward the center of the disk $\overline{K}$) at this point and, in addition, the inequality*

$$u(P') < u(P) \qquad (u(P') > u(P)),$$

where $u(P')$ is the value of u at the point P', holds true at any point $P' \in D$, then this derivative satisfies the inequality

$$u'_n(P) < 0 \qquad (u'_n(P) > 0).$$

Lemma 9.2 is a corollary of Lemma 1 from Sec. 28 in [35].

Lemma 9.3. *The difference of two solutions of Neumann's problem (9.1) is a constant.*

Proof.   It is evidently sufficient to prove that for $\psi_j(s) \equiv 0$, $1 \leq j \leq N$, problem (9.1) has no solutions other than constants. We assume that there is a function $w \not\equiv \text{const}$ on Ω that is a solution (in accordance with Definition 9.1) of a homogeneous Neumann's problem corresponding to problem (9.1). Then, by the maximum principle, $|w|$ attains its maximum, which we denote by $|w_0|$, at a point P_0 on the boundary of the polygon Ω, and the inequality

$$|w| < |w_0| \tag{9.3}$$

is satisfied on Ω.

Let us assume that the point P_0 does not coincide with any vertex of the polygon Ω, i.e., it lies in the interior of a side γ_{j_0}. Then, by virtue of (9.3) and Lemma 9.2, $w'_n \neq 0$ at the point P_0, and this contradicts the assumed homogeneity of the boundary conditions which the solution w satisfies. Therefore the point P_0 coincides with a vertex of the polygon Ω which we shall denote by $P_{j\bullet}$.

Suppose that $T_{j\bullet}$ is an extended sector (of radius $r_{j\bullet 0}$) of the covering of the polygon, corresponding to the vertex $P_{j\bullet}$, $w(r_{j^k}, \theta_{j^k})$ is the solution being considered. We carry out a conformal mapping

$$r = r_j^{1/\alpha_{j\bullet}}, \qquad \theta = \theta_{j\bullet}/\alpha_{j\bullet},$$

of the sector $T_{j\bullet}$ onto the half-disk T of radius $r_0 = r_{j\bullet 0}^{1/\alpha_{j\bullet}}$. This half-disk is supported by a diameter on which, except, maybe, the midpoint, the harmonic function

$$\tilde{w}(r, \theta) = w(r^{\alpha_{j\bullet}}, \alpha_{j\bullet}\theta) \tag{9.4}$$

satisfies a homogeneous boundary condition of the second kind. The possibility, resulting from this, of an (even) harmonic continuation of the function $\tilde{w}$ across the diameter with a delete point onto a disk with the deleted center indicates that the extended bounded function has a removable singularity at the deleted point. Therefore, by continuity, the function $\tilde{w}$ satisfies the homogeneous boundary condition of the second kind on the whole diameter. Hence, taking into account that $|\tilde{w}(0,0)| = |w_0|$ and that $|\tilde{w}| < |w_0|$ on T by virtue of (9.3) and (9.4), we arrive at a contradiction as before. We have proved Lemma 9.3.

Remark 9.1. Obviously, any constant is a solution of a homogeneous Neumann's problem.

Corollary 9.1. *If Neumann's problem* (9.1) *has a solution* v, *then the function*

$$V = v + C, \tag{9.5}$$

where C *is an arbitrary constant, is also a solution of problem* (9.1), *any solution of this problem being representable in form* (9.5) *for the appropriate choice of the constant* C.

Let us establish the existence of a solution of Neumann's problem. Along with Neumann's problem (9.1) we shall consider Dirichlet's problem

$$\Delta u = 0 \quad \text{on} \quad \Omega,$$
$$u = \varphi_j(s) \quad \text{on} \quad \gamma_j, \qquad j = 1, 2, \ldots, N, \tag{9.6}$$

where $\varphi_j(s)$ is an algebraic polynomial,

$$\varphi_j(s) = \varphi_{j-1}(s_j) + \int_{s_j}^{s} \psi_j(\sigma)\, d\sigma, \qquad s_j \leq s \leq s_{j+1}, \tag{9.7}$$

s_j is the length of the arc s corresponding to the end of the side γ_j at the vertex P_j, s_{N+1} is the value of s corresponding to the end of the side γ_N at the vertex P_1, $\varphi_0(s_1) = 0$.

By virtue of (9.2), the boundary values (9.7) are continuous at all vertices of the polygon Ω. Therefore, according to Theorem 1.1 (see also Definition 1.5), the solution u of Dirichlet's problem (9.6) is continuous up to the boundary of the polygon Ω, all the vertices inclusive, and the partial derivatives $\partial u/\partial x$ and $\partial u/\partial y$ are continuous up to the boundary, except for some vertices of the polygon Ω.

The harmonic function v, conjugate to the solution u of problem (9.6), related to u by the Cauchy–Riemann conditions

$$\frac{\partial u}{\partial x} = \frac{\partial v}{\partial y}, \qquad \frac{\partial u}{\partial y} = -\frac{\partial v}{\partial x},$$

is, since the polygon Ω is simply-connected, a single-valued function on this polygon up to an arbitrary additive constant (see [23]). It satisfies all requirements imposed on the solution of Neumann's problem (9.1) in accordance with Definition 9.1.

Let us verify this. Requirement (1) is obviously met. The continuous differentiability of the function v up to the boundary of the polygon Ω, except for its vertices, follows (see Remark 1.2) from possibility of an analytic continuation of the solution of problem (9.6) and, hence, of the function v across any side of the polygon Ω, except, maybe, its endpoints, i.e., requirement (3) is met. The continuity of the function v itself up to the vertices of the polygon Ω follows from its representation on the sector T_j, $1 \leq j \leq N$, in form (8.18), where $\mu = j$, $v_j = v$. Indeed, by virtue of the continuity of the boundary values (9.7) at the vertex P_j, $1 \leq j \leq N$, expression (8.1) does not contain the first term, and, consequently, the function S_j is continuous at the vertex P_j and the continuity at this point of the term, defined by the integral in (8.18), where $\mu = j$, is obvious. Thus requirement (2) is also met. Finally, the fulfillment of requirement (4) follows from (9.6), (9.7), Theorem 1.1 and the Cauchy-Riemann conditions according to which the derivative of the function u along the side γ_j (except for the endpoints) in the direction of its positive traverse coincides with the derivative of the function v along the inner normal to γ_j, $1 \leq j \leq N$.

Thus the function v, conjugate to the solution of Dirichlet's problem (9.6), is the solution of Neumann's problem (9.1) (in the sense of Definition 9.1).

The solvability of Neumann's problem on a multiply-connected polygon is investigated in [59].

9.2.   Approximate Solving of Neumann's Problem by Block Method

In order to find an approximate solution of Neumann's problem (9.1) we proceed as follows.

1. We seek the solution of the system of linear algebraic equations (4.8) corresponding to Dirichlet's problem (9.6).

2. On every closed basic block $\overline{T}^*_\mu$, $1 \le \mu \le M$, we define an approximation $v_{\mu n}$ for the function v conjugate to the solution u of problem (9.6) according to formula (8.17).

3. We piece together functions (8.17) as follows. We fix any $\mu = \overline{\mu}$ and set $C_{\overline{\mu}} = 0$. Next we consider all blocks $\overline{T}^*_\mu$ such that $\overline{T}^*_{\overline{\mu}} \cap \overline{T}^*_\mu \cap \Omega \ne \varnothing$ and choose the values of the arbitrary constants C_μ, corresponding to the indicated blocks, which satisfy the condition of coincidence of $v_{\mu n}$ with $v_{\overline{\mu} n}$ at any point of the set $\overline{T}^*_{\overline{\mu}} \cap \overline{T}^*_\mu$. And then we consider the blocks $\overline{T}^*_m$, which intersect in Ω at least one of the blocks $\overline{T}^*_\mu$, for which the constant C_μ is defined, and find C_m from the condition of coincidence of v_{mn} with any one of the completely defined functions $v_{\mu n}$ at the arbitrary point of the set $\overline{T}^*_\mu \cap \overline{T}^*_m (\overline{T}^*_\mu \cap \overline{T}^*_m \cap \Omega \ne \varnothing)$, etc. By virtue of condition IV (Sec. 2) imposed on the covering of the polygon Ω, the constants C_μ, $1 \le \mu \le M$ will be defined in some way. After this a common arbitrary constant C is added to all functions (8.17) ($1 \le \mu \le M$).

Lemma 9.4. *Suppose that $v_{\mu n}$ is an approximate solution of Neumann's problem (9.1) on the block $\overline{T}^*_\mu$ found by the method set forth above, $v_{\mu n}(P)$ is the value of $v_{\mu n}$ at the point $P \in \overline{T}^*_\mu$, v is the solution of Neumann's problem (9.1). Then for any points $P' \in \overline{T}^*_{\mu'}$, $P'' \in \overline{T}^*_{\mu''}$, $1 \le \mu' \le \mu'' \le M$,*

$$|(v_{\mu'' n}(P'') - v_{\mu' n}(P')) - (v(P'') - v(P'))| \le c^* \exp\{-dn\}, \quad (9.8)$$

where c^*, $d > 0$ are constants independent of the points P', P'', of the choice in which the constants, C_μ, $1 \le \mu \le M$, are chosen in (8.17), and of n for which system (4.8), corresponding to Dirichlet's problem (9.6), is uniquely solvable.

Lemma 9.4 is a corollary of Lemma 8.3 and a consequence of the continuity of the functions v and $v_{\mu n}$ up to the vertices P_μ of the polygon Ω, $1 \le \mu \le M$.

Remark 9.2. We can also use other methods to choose the constants C_μ, i.e., to splice together functions (8.17), $1 \le \mu \le M$. For instance, if we define one or several (independent of n) points at each of the nonempty intersections of different pairs of closed basic blocks, then the constants C_μ can be defined on the basis of the minimax approach to the deviations of the corresponding pairs of functions (8.17) at these points or by the method of the least squares. Inequality (9.8) remains valid, but maybe with another constant c^*.

Remark 9.3. Lemma 8.2 gives estimates for the errors of the derivatives of the function $v_{\mu n}$, $1 \leq \mu \leq M$.

For the solution of Neumann's problem by the block method on a multiply-connected polygon see [59].

10.   The Case of Arbitrary Analytic Mixed Boundary Conditions

Let us consider a mixed boundary-value problem of form (1.1), (1.2):

$$\Delta u = 0 \quad \text{on} \quad \Omega, \tag{10.1}$$

$$\nu_j u + \overline{\nu}_j u'_n = \varphi_j(s) \quad \text{on} \quad \gamma_j, \qquad j = 1, 2, \ldots, N, \tag{10.2}$$

and assume that $\varphi_j(s)$ are arbitrary real analytic functions on the sides γ_j of the polygon Ω, their endpoints inclusive. Here, as in problem (1.1), (1.2), ν_j is a given parameter equal to 0 or 1, $\overline{\nu}_j = 1 - \nu_j$, $1 \leq \nu_1 + \nu_2 + \ldots + \nu_N \leq N$, u'_n is a derivative along the inner normal to γ_j. We understand the solution of problem (10.1), (10.2) to be the function u obeying Definition 1.5 in items (4) and (5) of which the words "given polynomial" are replaced by the words "given function." Theorem 1.1 remains valid.

Let us touch upon the peculiarity of an approximate solution of problem (10.1), (10.2) by the block method and see what is in common with the case, studied in detail above, when the boundary conditions are defined by algebraic polynomials.

Conditions I–IV formulated in Sec. 2 must be satisfied by the covering of the polygon Ω by blocks of three types (sectors, half-disks, and disks). These conditions are explicitly connected only with the geometry of the polygon Ω. In this more general case of the arbitrary analytic functions $\varphi_j(s)$, $1 \leq j \leq N$, additional restrictions are imposed on the dimensions of the block-sectors and block-half-disks defined by these functions.

Let

$$\varphi_{j-1} = \sum_{k=0}^{\infty} a_{jk} r_j^k, \qquad \varphi_j = \sum_{k=0}^{\infty} b_{jk} r_j^k, \qquad (10.3)$$

where a_{jk}, b_{jk} are numerical coefficients, by the expansions in the power series of the functions φ_{j-1} and φ_j at the endpoints of the sides γ_{j-1} and γ_j at the vertex P_j of the polygon Ω. We denote by ρ_{j1} and ρ_{j0} the radii of convergence of series (10.3). We require that the radius r_{j0} of the extended block-sector T_j satisfy conditions I–III (Sec. 2) and the additional inequality

$$r_{j0} \leq \frac{2}{3} \min\{\rho_{j1}, \rho_{j0}\}, \qquad 1 \leq j \leq N. \qquad (10.4)$$

Furthermore, suppose that T_q is an extended block-half-disk the rectilinear part of whose boundary lies on the side $\gamma_{j(q)}$ of the polygon Ω (see Sec. 2), r_{q0} is the radius, and $P_q \in \gamma_{j(q)}$ is the center of the disk whose intersection with Ω gives a half-disk T_q, s_q is the length of the arc s corresponding to the point P_q, and, finally, let

$$\varphi_{j(q)} = \sum_{k=0}^{\infty} c_{qk}(s - s_q)^k, \qquad (10.5)$$

where c_{qk} are numerical coefficients, be the expansion in a power series of the function $\varphi_{j(q)}$ at the point $s = s_q$. We denote by ρ_q the radius of convergence of series (10.5). Let us require that along with conditions I–III (Sec. 2) the inequality

$$r_{q0} \leq 2\rho_q/3, \qquad N < q \leq L \qquad (10.6)$$

be also satisfied for the radius r_{q0}.

In the general case, restrictions (10.4), (10.6) imposed on the dimensions of the block-sectors and half-disks lead to an increase in the number of blocks in the covering of the polygon Ω. The covering evidently depends on the given functions φ_j, $1 \leq j \leq N$, via conditions (10.4), (10.6). It cannot be universal as in the case considered earlier, when the boundary conditions were defined only by algebraic polynomials and, consequently, the right-hand sides of (10.4), (10.6) go to infinity. The existence of a covering of the polynomial Ω by a finite number of blocks of three types, which satisfies conditions I–IV

(Sec. 2) and requirements (10.4), (10.6) for any analytic functions φ_j, $1 \le j \le N$, follows from the Heine-Borel lemma.

The existence of restrictions (10.4), (10.6) imposed on the dimensions of the extended block-sectors and half-disks guarantees a sufficiently rapid convergence of functional series which are calculated below when we seek an approximate solution of problem (10.1), (10.2).

After constructing the required covering of the polygon Ω by blocks, we define, as we did in Sec. 3, a harmonic carrier function $Q_j(r_j, \theta_j)$, satisfying the specified boundary conditions on the lateral sides of the extended block-sector T_j in the neighborhood of every vertex P_j. Dependent on the values of the parameters ν_{j-1} and ν_j the carrier function is defined as follows:

(1) for $\nu_{j-1} = \nu_j = 1$ we have

$$Q_j(r_j, \theta_j) = b_{j0} + \frac{a_{j0} - b_{j0}}{\alpha_j \pi} \theta_j$$

$$+ \sum_{k=1}^{\infty} (a_{jk} \xi_{jk}(r_j, \theta_j) + b_{jk} \xi_{jk}(r_j, \alpha_j \pi - \theta_j)); \qquad (10.7)$$

(2) for $\nu_{j-1} = \nu_j = 0$ we have

$$Q_j(r_j, \theta_j) = \sum_{k=0}^{\infty} (a_{jk} \eta_{jk}(r_j, \theta_j) + b_{jk} \eta_{jk}(r_j, \alpha_j \pi - \theta_j)); \qquad (10.8)$$

(3) for $\nu_{j-1} = 0$, $\nu_j = 1$ we have

$$Q_j(r_j, \theta_j) = \sum_{k=0}^{\infty} (a_{jk} \vartheta_{jk}(r_j, \theta_j) + b_{jk} \varkappa_{jk}(r_j, \alpha_j \pi - \theta_j)); \qquad (10.9)$$

(4) for $\nu_{j-1} = 1$, $\nu_j = 0$ we have

$$Q_j(r_j, \theta_j) = \sum_{k=0}^{\infty} (a_{jk} \varkappa_{jk}(r_j, \theta_j) + b_{jk} \vartheta_{jk}(r_j, \alpha_j \pi - \theta_j)), \qquad (10.10)$$

where ξ_{jk}, η_{jk}, ϑ_{jk}, $\varkappa_{jk}$ are functions (3.3), (3.5), (3.7), (3.8), $\alpha_j \pi$ is the magnitude of the angle of the polygon Ω in which the block-sector T_j lies, a_{jk}, b_{jk} are the same numerical coefficients as those of series (10.3).

The functional series in (10.7)–(10.10) converge uniformly on the sector

$$T_{j0} = \{(r_j, \theta_j) : 0 < r_j < \frac{6}{5} r_{j0}, -\alpha_j \pi < \theta_j < 2\alpha_j \pi\}, \qquad (10.11)$$

lying, in general, on a Riemannian surface with a branch point $r_j = 0$. This sector contains an extended block-sector T_j. The sum of each of the indicated series is a function bounded and harmonic on $T_{j0} \supset T_j$. The series admits a termwise differentiation with respect to x and y on the sector $T_{j0} \supset (\overline{T}_j \setminus P_j)$ any number of times. Moreover, the series termwise differentiated with respect to x and y a finite number of times converge on $\overline{T}_j \setminus P_j$ in a uniform metric in a geometrical progression to the corresponding derivatives of the sums of these series. Finally, we must point out that for the values of the parameters ν_{j-1} and ν_j chosen in (10.2) the function $Q_j(r_j, \theta_j)$ satisfies the boundary conditions defined in (10.2) on $(\gamma_{j-1} \cap T_{j0}) \setminus P_{j-1}$ and on $(\gamma_j \cap T_{j0}) \setminus P_{j+1}$.

Let us prove these assertions for $\nu_{j-1} = \nu_j = 1$. We set

$$Q_j^1(r_j, \theta_j) = \sum_{k=1}^{\infty} a_{jk} \xi_{jk}(r_j, \theta_j) \qquad (10.12)$$

and estimate the terms of this series on the closed sector $\overline{T}_{j0}$. It is evident from (3.3) that for all $k \geq 1$, for which $\sin k\alpha_j \pi = 0$ or $|\sin k\alpha_j \pi| \geq 1/2$ there is a constant c_j^0 independent of k, such that the inequality

$$|a_{jk} \xi_{jk}(r_j, \theta_j)| \leq c_j^0 |a_{jk}| r_j^{k-1} \quad \text{on} \quad \overline{T}_{j0} \qquad (10.13)$$

holds true.

We shall consider the remaining case when

$$0 < |\sin k\alpha_j \pi| < 1/2, \qquad (10.14)$$

in greater detail. We note, first of all, that by virtue of (10.14)

$$|\sin k\alpha_j \pi| = \sin \omega_{jk} \pi, \qquad (10.15)$$

where

$$\omega_{jk} = \min\{k\alpha_j - [k\alpha_j], 1 + [k\alpha_j] - k\alpha_j\}, \qquad (10.16)$$

with

$$0 < \omega_{jk} < 1/6. \qquad (10.17)$$

Inasmuch as $k' = [k\alpha_j + 1/2]/\alpha_j \geq 0$ (see (3.3)), we have

$$|k' - k| = \frac{|1/2 - \{k\alpha_j + 1/2\}|}{\alpha_j} = \frac{\omega_{jk}}{\alpha_j}, \qquad (10.18)$$

where $\{\ \}$ is the sign of the fractional part, under conditions (10.14).
We assume that

$$|k' - k| \geq 1/2, \qquad k \geq 1. \qquad (10.19)$$

By virtue of (10.15)–(10.18) this is possible only when $\omega_{jk} \geq \alpha_j/2$, $0 < \alpha_j < 1/3$ and, consequently,

$$|\sin k\alpha_j \pi| \geq \sin(\alpha_j \pi/2) > 0,$$

where α_j does not depend on k. From this result and from (3.3), with due regard for (10.18), (10.17) we get inequality (10.13) in the subcase (10.19), but maybe with another constant c_j^0 independent of k.

It remains to consider a subcase when

$$0 < |k' - k| < 1/2, \qquad k \geq 1. \qquad (10.20)$$

Under conditions (10.14), (10.20) we easily find from (3.3) that

$$|a_{jk}\xi_{jk}(r_j, \theta_j)| \leq |a_{jk}|r_j^{k-1}\left(r_j \left|\frac{\sin k\theta_j - \sin k'\theta_j}{\sin k\alpha_j \pi}\right|\right)$$

$$+ |a_{jk}|r_j^{k-1}(1 + (5/6r_{j0})^{k'-k}) \times \begin{cases} r_j\dfrac{1 - (5r_j/6r_{j0})^{k'-k}}{|\sin k\alpha_j \pi|}, & k' > k, \\[2ex] r_j^{1+k'-k}\dfrac{1 - (5r_j/6r_{j0})^{k-k'}}{|\sin k\alpha_j \pi|}, & k' < k, \end{cases} \qquad (10.21)$$

where $0 \leq r_j \leq 6r_{j0}/5$, $1 + k' - k > 1/2$.

By virtue of (10.15)–(10.18) we have

$$|\sin k\theta_j - \sin k'\theta_j| \le 2\left|\sin \frac{\omega_{jk}\theta_j}{2\alpha_j}\right| \le 2|\sin k\alpha_j\pi|,$$

$$-\alpha_j \le \theta_j \le 2\alpha_j\pi. \tag{10.22}$$

Moreover, according to (10.15)–(10.18) we have

$$|\sin k\alpha_j\pi| = \sin \omega_{jk}\pi > 3\omega_{jk} = 3\alpha_j|k' - k|. \tag{10.23}$$

Proceeding from (10.21)–(10.23), with due regard for (3.3), (10.11), and the elementary estimate

$$\frac{1 - a^\varepsilon}{\varepsilon} \le \ln \frac{1}{a}, \qquad 0 < a \le 1, \qquad 0 < \varepsilon \le 1,$$

we arrive at inequality (10.13) with a certain constant c_j^0 independent of k for subcase (10.20).

We assume now that inequality (10.13) includes the constant c_j^0 which is maximal in all the cases and subcases considered. Inequality (10.13) is valid for any $k \ge 1$.

It follows from inequality (10.4) that the number series

$$\sum_{k=1}^{\infty} |a_{jk}|(7r_{j0}/5)^{k-1}$$

is convergent and, consequently, the sequence

$$|a_{jk}|(7r_{j0}/5)^{k-1}, \qquad k = 1, 2, \dots,$$

is bounded. This property and estimate (10.13) imply the convergence of the functional series (10.12) in a uniform metric in a geometric progression with a common ratio $q = 6/7 < 1$ on the sector T_{j0} up to its boundary. Since the terms of series (10.12) are harmonic on the sector T_{j0} and continuous up to its boundary, the sum of this series is also continuous on the sector T_{j0} up to its boundary and harmonic on the sector $T_{j0} \supset \overline{T}_j \setminus P_j$ according to Weierstrasse's theorem.

We can establish by analogy that the series

$$Q_j^2(r_j,(\theta_j)) = \sum_{k=1}^{\infty} b_{jk}\xi_{jk}(r_j,\alpha_j\pi - \theta_j) \qquad (10.24)$$

converges in a uniform metric in a geometric progression on the sector T_{j0} up to its boundary and the sum of this series is harmonic on T_{j0} and continuous up to the boundary of the sector T_{j0}.

Thus function (10.7) is bounded and harmonic on the sector $T_{j0} \supset T_j$ and continuous up to its boundary, except for the vertex P_j, if $a_{j0} \neq b_{j0}$.

It is obvious that

$$Q_j^1(r_j,0) = 0, \qquad Q_j^1(r_j,\alpha_j\pi) = \sum_{k=1}^{\infty} a_{jk}r_j^k,$$

$$Q_j^2(r_j,0) = \sum_{k=1}^{\infty} b_{jk}r_j^k, \qquad Q_j^2(r_j,\alpha_j\pi) = 0.$$

Consequently, the carrier function (10.7) satisfies the boundary conditions defined in (10.2), where φ_{j-1}, φ_j are represented as (10.3), for $\nu_{j-1} = \nu_j = 1$ on $(\gamma_{j-1} \setminus P_{j-1}) \cap T_{j0}$ and on $(\gamma_j \setminus P_{j+1}) \cap T_{j0}$.

By virtue of the uniform convergence of the functional series in (10.7) on T_{j0} and of Lemma 3.4, this series admits of a termwise differentiation with respect to x and y any number of times on the sector T_{j0} (10.11). Evidently, there is a constant $c_j > 0$ such that any point $(r_j,\theta_j) \in \overline{T}_j \setminus P_j$ can be placed at center of a disk of radius $c_j r_j$ which entirely lies in the sector T_{j0} (10.11). Therefore, by virtue of estimate (10.13) for the first term and a similar estimate for the second term in the expression for the terms of the series in (10.7) and of Lemma 3.4, we have

$$\left| \frac{\partial^p (a_{jk}\xi_{jk}(r_j,\theta_j) + b_{jk}\xi_{jk}(r_j,\alpha_j\pi - \theta_j))}{\partial x^{p-q}\partial y^q} \right|$$

$$\leq c_{jp}(|a_{jk}| + |b_{jk}|)r_j^{k-1-p} \quad \text{on} \quad \overline{T}_j \setminus P_j,$$

where T_j is an extended block-sector, $p = 0,1,\ldots$, $0 \leq q \leq p$, c_{jp} is a constant independent of r_j, θ_j and $k \geq 1 + p$. With due account of (10.4) it easily follows from this inequality that the functional series in (10.7) admits of a termwise differentiation with respect to x and y on $\overline{T}_j \setminus P_j$ any number of times, the differentiated series being convergent on $\overline{T}_j \setminus P_j$ in a uniform metric in a geometric progression.

We can establish by analogy that functions (10.8)–(10.10) are harmonic on the sector T_{j0} (10.11) and are continuous up to its boundary, and series (10.8)–(10.10) admit of a termwise differentiation with respect to x and y on T_{j0} any number of times. Moreover, the series differentiated term-by-term converge in a geometric progression on $\overline{T}_j \setminus P_j$, where T_j is an extended block-sector.

With due regard for (10.3), direct verifications show that for the values of the parameters ν_{j-1} and ν_j chosen in (10.2) one of the functions (10.7)–(10.10) corresponding to them satisfies the specified boundary conditions on $(\gamma_{j-1} \setminus P_{j-1}) \cap T_{j0}$ and on $(\gamma_j \setminus P_{j+1}) \cap T_{j0}$.

In addition to the carrier function $Q_j(r_j, \theta_j)$, defined in the neighborhoods of the vertices P_j, $1 \leq j \leq N$, of the polygon Ω, the harmonic functions $Q_q(r_q, \theta_q)$, $N < q \leq L$, are defined on the disks

$$T_{q0} = \{(r_q, \theta_q) : 0 \leq r < 6r_{q0}/5, \, 0 \leq \theta_q < 2\pi\}, \qquad (10.25)$$

containing the corresponding closed extended block-half-disks $\overline{T}_q$ (see (2.1), (2.3)).

The function $Q_q(r_q, \theta_q)$, $N < q \leq L$, has the form

$$Q_q(r_q, \theta_q) = \sum_{k=0}^{\infty} c_{qk} r_q^k \left(\nu_{j(q)} \cos k\theta_q + \overline{\nu}_{j(q)} r_q \frac{\sin(k+1)\theta_q}{k+1} \right) \qquad (10.26)$$

and, with (10.5), (10.6), (10.25) taken into account, satisfies the boundary condition defined in (10.2) on

$$(\gamma_{j(q)} \setminus (P_{j(q)} \cup P_{j(q)+1})) \cap T_{q0},$$

where $\gamma_{j(q)}$ is the side on which the rectilinear part of the boundary of the block-half-disk T_q lies. On the basis of (10.6), (10.25), series (10.26) uniformly converges on the closed disk $\overline{T}_{q0}$. Hence the continuity of function (10.26) on $\overline{T}_{q0}$ and its harmonicity on the open disk $T_{q0} \supset \overline{T}_q$, and, with due regard for Lemma 3.4 the possibility of a termwise differentiation of series (10.26) with respect to x and

y on T_{q0} any number of times. In this case the termwise differentiated series (10.26) converges on the closed extended block-half-disk
$\overline{T}_q \subset T_{q0}$ in a geometric progression by virtue of (10.6).

In the general case, besides block-sectors and half-disks the covering of the polygon Ω also includes block-disks T_p, T_p^* (see (2.2),
(2.3)), $L < p \leq M$. As before, we use formula (3.11) to determine
the functions $Q_p(r_p, \theta_p)$, $L < p \leq M$.

We have thus constructed the carrier functions $Q_\mu(r_\mu, \theta_\mu)$ for all
extended blocks T_μ, $1 \leq \mu \leq M$. The function Q_μ is bounded and
harmonic on the open extended block T_μ and continuous up to its
boundary, except for the point P_μ, if the block T_μ is a sector, and
the boundary conditions defined in (10.2) for $\nu_{\mu-1} = \nu_\mu = 1$ are
discontinuous at its vertex P_μ. In addition, these functions satisfy
the boundary conditions defined in (10.2) on the rectilinear parts of
the boundaries of the block-half-disks and sectors, except, maybe, the
vertices of the latter.

When the carrier functions Q_μ, $1 \leq \mu \leq M$, possess the indicated
properties, Theorem 3.1 on the integral representation of the solution of a mixed boundary-value problem on blocks can be evidently
extended to the case of boundary conditions analytic on the sides of
a polygon, i.e., to the boundary-value problem (10.1), (10.2).

As before, we can find the approximate solution of this boundary
value problem on the closed basic blocks $\overline{T}_\mu^*$, $1 \leq \mu \leq M$, in form
(5.1), where u_μ^k, $1 \leq \mu \leq M$, $1 \leq k \leq n_\mu$, is a solution of the
corresponding system of linear algebraic equations (4.8). The only
difference is that the carrier functions Q_μ, $1 \leq \mu \leq L$, are defined not
as a finite sum of elementary functions as it was done in Sec. 3, but,
generally speaking, by means of series (10.7)–(10.10), (10.26).

Theorem 5.1 on the exponential convergence of an approximate
solution remains completely valid under boundary conditions analytic
on the sides of the polygon Ω. Its proof given in Sec. 6 does not
essentially change when the functions Q_μ, $1 \leq \mu \leq M$, that possess all
the properties used in the proof, are defined by the method described
above. In particular, the carrier function Q_μ, $1 \leq \mu \leq L$, is harmonic
and bounded on a certain domain which contains the curvilinear part
of the boundary of the extended block $T_\mu \subset T_{\mu 0}$ strictly in its interior.

This property is used in the proof of Lemma 6.10.

The investigations carried out in Sec. 7 and the corresponding assertions connected with the stability of the block method can evidently be extended without any changes to the boundary-value problem (10.1), (10.2). As concerns the estimates of the number of the needed arithmetic operations given at the end of Sec. 7, they must be improved in the case being considered. First, the order of the estimate of the number N_2 of operations needed to calculate the coefficients $\tilde{R}_{\tau\mu}^{km}$, $\tilde{a}_\mu^k$ is higher when the carrier functions $Q_q(r_q, \theta_q)$, $1 \leq q \leq L$, are defined by series (10.7)–(10.10), (10.26), namely,

$$N_2 = O(|\ln^3 \varepsilon|).$$

However, this does not change the total estimate (7.42) of the number of the arithmetic operations needed to find the approximate solution of the corresponding algebraic problem (4.8). Second, when the values are stored as was indicated in Remark 7.8, the number of arithmetic operations needed to calculate one value of the approximate solution (7.30) with an accuracy to within $O(\varepsilon)$ is estimated by the value $O(\ln^2 \varepsilon)$ (rather than $O(|\ln \varepsilon|)$). The first term in (7.30) is defined by a series of the indicated kind, convergent on $\overline{T}_\mu^*$ in a uniform metric in a geometric progression.

Remark 10.1. It is of interest to consider a special case when boundary conditions are defined and closed polygonal lines that form the boundary of the polygon Ω by the traces of elementary functions which are harmonic in some neighborhoods of the corresponding polygonal lines. In the general case, on different polygonal lines we take the traces of different harmonic functions which may not be defined throughout the polygon Ω. In this case, the carrier functions Q_μ coincide, on extended block-sectors and half-disks adjoining the fixed polygon line, with an elementary harmonic function whose trace defines the boundary conditions on this polygonal line (see, in particular, Sec. 11 and Sec. 13 below).

Chapter 2

Approximate Block Method of Conformal Mapping of Polygons onto Canonical Domains

11. Approximate Conformal Mapping of a Simply-Connected Polygon onto a Disk

Suppose that in the complex plane z, $z = x + iy$, there is an open simply-connected polygon Ω (see Sec. 1), to be conformally mapped onto a unit disk $|\zeta| < 1$, $\zeta = \xi + i\eta$, so that a point z_0 belonging to Ω passes into the center of the disk. The required mapping function $\zeta = f(z)$ is known (see [23, item 43]) to be representable in the form

$$f(z) = (z - z_0) \exp\{u(z) + iv(z) + i\beta\}, \qquad (11.1)$$

where u is a solution of Dirichlet's problem

$$\begin{aligned} \Delta u &= 0 \quad \text{on} \quad \Omega, \\ u &= \ln|z - z_0|^{-1} \quad \text{on} \quad \gamma_j, \qquad j = 1, 2, \ldots, N, \end{aligned} \qquad (11.2)$$

v is a harmonic function conjugate to u, and β is an arbitrary constant.

We shall seek approximations of the functions u and v, for which purpose we specify a covering of the polygon Ω by a finite number of blocks according to the rules set forth in Sec. 2. In addition we require that the closed extended blocks $\overline{T}_q$, $1 \leq q \leq L$, that are sectors and half-disks, should not contain the point $z_0 \in \Omega$. It is important in practice that the distance between the point z_0 and the set $\overline{T}_1 \cup \overline{T}_2 \cup \ldots \overline{T}_L$ not be too small. Thus the covering must contain at least one basic block $\overline{T}_p^*$ as a disk on which the point z_0 lies.

We set

$$Q_q(r_q, \theta_q) = \ln |z - z_0|^{-1} \quad \text{on} \quad \overline{T}_q, \tag{11.3}$$

$$S_q(r_q, \theta_q) = -\arg (z - z_0) \quad \text{on} \quad \overline{T}_q, \tag{11.4}$$

where $1 \leq q \leq L$, $z = z(r_q, \theta_q) = x(r_q, \theta_q) + iy(r_q, \theta_q)$, with $\arg (z - z_0)$ chosen so as to have a branch that is single-valued and continuous on $\overline{T}_q$,

$$Q_p(r_p, \theta_p) = S_p(r_p, \theta_p) = 0 \quad \text{on} \quad \overline{T}_p, \tag{11.5}$$

$$L < p \leq M.$$

Suppose that u_μ^k, $1 \leq \mu \leq M$, $1 \leq k \leq n_\mu$, is a solution of system (4.8), and $u_{\mu n}$, $v_{\mu n}$ are functions (5.1), (8.17), corresponding to problem (11.2) and, respectively, to the functions Q_μ, S_μ defined by formulas (11.3)–(11.5). We piece functions (8.17) together in the way described in Sec. 9 (item 3) with the only difference that we choose $\overline{\mu} = 1$ and, besides, do not add the common arbitrary constants after piecing the functions together.

Theorem 11.1. *Suppose that the function $\zeta = f(z)$ conformally maps the polygon Ω onto a disk $|\zeta| < 1$ with $f(z_0) = 0$, $f(z_1) = 1$, where z_1 is a complex coordinate of the vertex P_1 of the polygon Ω. Then*

$$|f_{\mu n}(z) - f(z)| \leq c_0 \exp\{-d_0 n\} \quad \text{on} \quad \overline{T}_\mu^*, \tag{11.6}$$

where $\overline{T}_\mu^*$ is a closed basic block,

$$f_{\mu n}(z) = (z - z_0) \exp\{u_{\mu n} + iv_{\mu n}\}, \tag{11.7}$$

$1 \leq \mu \leq M$, c_0, $d_0 > 0$ *are constants independent of the choice of the points at which the functions $v_{\mu n}$ are pieced together and of n for which system (4.8) is uniquely solvable.*

Theorem 11.1 follows from (11.1), (11.7) on the basis of Theorem 5.1 and Lemmas 8.1–8.3 which can be extended to Problem (11.2) under the condition that the covering of the polygon Ω satisfies the additional requirement indicated above, and the functions Q_μ and S_μ are defined by formulas (11.3)–(11.5). The proofs of Theorem 5.1 and Lemmas 8.1–8.3 can be carried over to problem (11.2) without any changes.

Theorem 6.1 guarantees the unique solvability of system (4.8) for a sufficiently large n.

Remark 11.1. This method can be used to find the approximate expression (11.7) for the function $f(z)$ on every closed basic block $\overline{T}_\mu^* \subset \overline{\Omega}$, $1 \leq \mu \leq M$, taken separately. This expression is an analytic (elementary) function on the blocks, except, maybe, the vertices of the block-sectors. If $\overline{T}_\mu^* \cap \overline{T}_\nu^* \cap \Omega \neq \emptyset$, then there are at least two approximate expressions for $f(z)$ at the intersection of the blocks $\overline{T}_\mu^*$ and $\overline{T}_\nu^*$ which are close to $f(z)$ according to Theorem 11.1 and, consequently, to each other.

Remark 11.2. The Schwars–Christoffel formula (see [23]), into which we have substituted the approximate images of the vertices of the polygon Ω found earlier by the block method, yields an approximate conformal mapping of a unit disk on a polygon, similar to Ω, with a uniform accuracy $O(\exp\{-d_0 n\})$, where $d_0 > 0$ is the same constant as in (11.6).

Remark 11.3. For homogeneous boundary conditions of the first kind Green's function of Laplace's operator can be expressed in terms of function (11.1) as follows (see [19, Ch. 5, Sec. 8])

$$G(z; z^0) = \frac{1}{2\pi} \ln \frac{|1 - \overline{\zeta}^0 f(z)|}{|f(z) - \zeta^0|}, \tag{11.8}$$

where $\zeta^0 = f(z^0)$, $z^0 \in \Omega$. Replacing the function $f(z)$ in (11.8) by its approximation (11.7) and setting $\zeta^0 = f_{\tau n}(z^0)$, where τ is the minimal index of the closed basic block $\overline{T}_\tau^*$ containing the point z^0, we get an approximation for Green's function on the closed basic block $\overline{T}_\mu^*$, $1 \leq \mu \leq M$.

12. Basic Harmonic Functions

Suppose we are given a t-connected polygon Ω, $t \geq 2$. We denote the closed polygonal lines that form the boundary of the polygon Ω by Γ_l, $l = 1, 2, \ldots, t$, and understand Γ_1 to be its external boundary. Then we number the vertices and sides of the polygon Ω according to the rules presented in Sec. 1 passing from one polygonal line to another in the order of increasing of their indices. We denote by N_l the largest index of the side $\gamma_j \in \Gamma_l$ and set $N_0 = 0$, $N = N_t$.

Let us consider Dirichlet's problem

$$\begin{aligned} \Delta u^\lambda &= 0 \quad \text{on} \quad \Omega, \\ u^\lambda &= \delta_l^\lambda \quad \text{on} \quad \Gamma_l, \quad l = 1, 2, \ldots, t, \end{aligned} \tag{12.1}$$

where δ_l^λ is Kronecker's delta, $\lambda = 1, 2, \ldots, t$.

We shall call the solutions u^λ, $\lambda = 1, 2, \ldots, t$, of problem (12.1) basic harmonic functions on the polygon Ω. We use them in the subsequent sections to construct conformal mapping of multiply-connected polygons.

We set $1 \leq l, \lambda \leq t$,

$$\varkappa_l^\lambda = -\int_{\Gamma_l} (u^\lambda)'_n \, ds \equiv - \sum_{j=N_{l-1}+1}^{N_l} \int_{\gamma_j} (u^\lambda)'_n \, ds. \tag{12.2}$$

Lemma 12.1. *The relations*

$$\sum_{l=1}^{t} \varkappa_l^\lambda = 0, \qquad 1 \leq \lambda \leq t, \tag{12.3}$$

$$\sum_{\lambda=1}^{t} \varkappa_l^\lambda = 0, \qquad 1 \leq l \leq t, \tag{12.4}$$

$$\varkappa_\lambda^\lambda > 0, \qquad \varkappa_l^\lambda < 0, \qquad l \neq \lambda, \qquad 1 \leq l, \lambda \leq t, \tag{12.5}$$

hold true.

Proof. Relations (12.3) follows from Lemma 3.3. According to the principle of maximum we have $0 < u^\lambda < 1$ on Ω, $1 \leq \lambda \leq t$. Consequently, inequalities (12.5) are satisfied by virtue of Lemma 9.2. Relations (12.4) are a consequence of the identity $u^1 + u^2 + \ldots + u^t \equiv 1$ on $\overline{\Omega}$. We have proved Lemma 12.1

Lemma 12.2. *The determinant of the matrix* $\|\varkappa_i^\lambda\|_{\lambda,l=2}^t$ *is non-zero.*

Lemma 12.2 follows from Lemma 12.1 by which the matrix $\|\varkappa_i^\lambda\|_{\lambda,l=2}^t$ has a dominant principal diagonal.

Remark 12.1. With due account of the rate of growth of the partial derivatives of the solutions of problem (12.1) in the vicinity of the vertices of the polygon Ω established by inequality (3.21), it follows from the Cauchy–Riemann conditions that quantity (12.2) is the increment of the harmonic function conjugate to u^λ under the positive traverse of the polygonal line Γ_l.

Lemma 12.3. *Every function harmonic on a t-connected polygon Ω, that assumes constant values on the polygonal lines Γ_l, $l = 1, 2, \ldots, t$, forming its boundary, has a single-valued conjugate (to within an additive constant) is a constant.*

Proof.   For $t = 1$ the statement of Lemma 12.3 is trivial. Suppose that the function u possesses specified properties on the t-connected polygon Ω, $t \geq 2$. Without loss of generality, we assume that $u = 0$ on Γ_1. Obviously, the function u is representable in the form

$$u = \sum_{\lambda=2}^{t} c_{0\lambda} u^\lambda,$$

where $c_{0\lambda}$ is the value the function u assumes on Γ_λ, u^λ is the solution of problem (12.1).

By the hypothesis, the increment of the function conjugate to u under the traverse Γ_l, $l = 2, 3, \ldots, t$, is zero, i.e.,

$$\sum_{\lambda=2}^{t} c_{0\lambda} \varkappa_i^\lambda = 0, \qquad l = 2, 3, \ldots, t.$$

Hence $c_{0\lambda} = 0$, $\lambda = 2, 3, \ldots, t$, according to Lemma 12.2. Lemma 12.3 is proved.

13.   Approximate Conformal Mapping of a Multiply-Connected Polygon onto a Plane with Cuts along Parallel Line Segments

13.1.   Structure of a Mapping

According to Hilbert's theorem [17, Ch. 5, Sec. 2, Theorem 1] there is a conformal mapping of a t-connected polygon Ω lying in the complex plane z onto a complex plane ζ with cuts along t segments parallel to the imaginary axis. The mapping function can be written as

$$\zeta = 1/(z - z_0) + \zeta_0(z), \tag{13.1}$$

where $z = x + iy$, $z_0 = x_0 + iy_0$ ($z_0 \in \Omega$) is fixed, and $\zeta_0(z)$ is analytic in Ω. Under the additional requirement

$$\zeta_0(z_0) = 0 \tag{13.2}$$

a conformal mapping of form (13.1) is unique [17, Ch. 5, Sec. 2, Theorem 2].

The real part of the function ζ assumes a constant value on each polygonal line Γ_l forming part of the boundary of the polygon Ω, i.e.,

$$\mathrm{Re}\,\zeta = c_l \quad \text{on} \quad \Gamma_l, \qquad l = 1, 2, \ldots, t, \tag{13.3}$$

the constants c_l being unknown beforehand. Let us find them. We denote by $\delta_l f(z)$ the increment of the function $f(z)$ under the traverse of the polygonal line Γ_l in the positive direction (when Ω is locally on the left) and set

$$u = \mathrm{Re}\,\zeta_0, \qquad v = \mathrm{Im}\,\zeta_0. \tag{13.4}$$

On the basis of (13.1) and (13.3) we have

$$u = c_l - \mathrm{Re}\,\frac{1}{z - z_0} \quad \text{on} \quad \Gamma_l, \qquad l = 1, 2, \ldots, t, \tag{13.5}$$

and, since the function $\zeta_0(z)$ is analytic in Ω,

$$\delta_l v = 0, \qquad l = 1, 2, \ldots, t. \tag{13.6}$$

Let u^0 be the solution of the boundary-value problem

$$\Delta u^0 = 0 \quad \text{on} \quad \Omega,$$

$$u^0 = -\operatorname{Re}(z - z_0)^{-1} \quad \text{on} \quad \Gamma_l, \qquad l = 1, 2, \ldots, t. \qquad (13.7)$$

Then the function u can obviously be represented as

$$u = c_1 + u^0 + \sum_{\lambda=2}^{t} c'_\lambda u^\lambda, \qquad (13.8)$$

where u^λ is the solution of the boundary-value problem (12.1),

$$c'_\lambda = c_\lambda - c_1, \qquad \lambda = 2, 3, \ldots, t, \qquad (13.9)$$

and c_l, $l = 1, 2, \ldots, t$, are the required constants.

We denote by $\varkappa_l^0$ the quantity defined by (12.2) for $\lambda = 0$ (via the solution of problem (13.7)). We require that in accordance with (13.6) an increment of the function v conjugate to u be zero under the traverse of the polygonal lines Γ_l, $l = 2, 3, \ldots, t$, i.e., we subject the constants c'_λ to the conditions

$$\varkappa_l^0 + \sum_{\lambda=2}^{t} c'_\lambda \varkappa_l^\lambda = 0, \qquad l = 2, 3, \ldots, t. \qquad (13.10)$$

By virtue of Lemma 12.2, the constants c'_λ, $\lambda = 2, 3, \ldots, t$, are uniquely defined by system (13.10). Finally we set

$$c_1 = -u^0(z_0) - \sum_{\lambda=2}^{t} c'_\lambda u^\lambda(z_0). \qquad (13.11)$$

Thus, for the constants c_1, c'_2, c'_3, $\ldots$, c'_t function (13.8) harmonic on Ω satisfies the condition $u(z_0) = 0$ and has on Ω a single-valued conjugate function v such that $v(z_0) = 0$. The function

$$\operatorname{Re} \frac{1}{z - z_0} + u$$

obviously assumes constant values c_l on the polygonal lines Γ_l, $l = 1, 2, \ldots, t$, and, moreover, c_1 has expression (13.11) and $c_l = c'_l + c_1$, $l = 2, 3, \ldots, t$ according to (13.9).

By virtue of Lemma 12.3, the functions u and v we have found, i.e., functions (13.4), are unique.

13.2. Constructing an Approximate Mapping

The method of finding the function $\zeta_0(z)$ can be realized approximately as follows.

1. A covering of the given t-connected polygon Ω is constructed according to the rules presented in Sec. 2 with an additional condition, similar to that given in Sec. 11, that the closed extended block-sectors and block-half-disks $\overline{T}_q$, $1 \leq q \leq L$, not contain the point z_0. We assume that the sides γ_j of the polygon Ω are labeled in the same way as in Sec. 12.

Let Ξ_l be the set of indices of the block-sectors and block-half-disks the rectilinear parts of whose boundaries lie on Γ_l, i.e.,

$$\Xi_l = \{\mu : P_\mu \in \Gamma_l, 1 \leq \mu \leq L\},$$

and, in addition, $\Xi = \Xi_1 \cup \Xi_2 \cup \ldots \cup \Xi_t$.

2. We define the functions

$$Q_{\mu 0}(z) = \begin{cases} -\operatorname{Re}(z - z_0)^{-1}, & \mu \in \Xi, \\ 0, & \mu \notin \Xi, \end{cases}$$

$$S_{\mu 0}(z) = \begin{cases} -\operatorname{Im}(z - z_0)^{-1}, & \mu \in \Xi, \\ 0, & \mu \notin \Xi, \end{cases}$$

$$Q_{\mu \lambda}(z) = \begin{cases} 1, & \mu \in \Xi_\lambda, \\ 0, & \mu \notin \Xi_\lambda, \end{cases}$$

$\lambda = 2, 3, \ldots, t$, and the kernels

$$\mathop{R_p}_{I_p}(r_p, \theta_p, \eta) = \mathop{R}_{I}\left(\frac{r_p}{r_{p0}}, \theta_p, \eta\right),$$

$$\mathop{R_q}_{I_q}(r_q, \theta_q, \eta) = \sum_{k=0}^{1}(-1)^k \mathop{R}_{I}\left(\frac{r_q}{r_{q0}}, \theta_q, (-1)^k \eta\right),$$

$$\mathop{R_j}_{I_j}(r_j, \theta_j, \eta) = \frac{1}{\alpha_j}\sum_{k=0}^{1}(-1)^k \mathop{R}_{I}\left(\left(\frac{r_j}{r_{j0}}\right)^{1/\alpha_j}, \frac{\theta_j}{\alpha_j}, (-1)^k \frac{\eta}{\alpha_j}\right),$$

corresponding to the boundary-value problem (13.7), where $1 \leq j \leq N < q \leq L < p \leq M$, R, I are functions (3.14), (8.13). The expressions for these kernels are simpler than those in cases (3.16), (3.17), (8.15), (8.16), since they are related only to Dirichlet's problem.

3. For every $\lambda = 0, 2, 3, \ldots, t$ we find a solution $u_\mu^{k\lambda}$, $1 \leq \mu \leq M$, $1 \leq k \leq n_\mu$, of the system of linear algebraic equations

$$u_\mu^{k\lambda} = Q_{\tau\mu}^{k\lambda} + \beta_\tau \sum_{m=1}^{n_\tau} (u_\tau^{m\lambda} - Q_\tau^{m\lambda}) R_{\tau\mu}^{km}, \qquad (13.12)$$

$$1 \leq \mu \leq M, \qquad 1 \leq k \leq n_\mu, \qquad \tau = \tau(\mu, k),$$

of form (4.8) corresponding to problem (13.7) for $\lambda = 0$ and to problem (12.1) for $\lambda = 2, 3, \ldots, t$. Here $R_{\tau\mu}^{km}$, β_τ are quantities (4.7), (4.2),

$$Q_{\tau\mu}^{k\lambda} = Q_{\tau\lambda}(z_\mu^k), \qquad Q_\tau^{m\lambda} = Q_{\tau\lambda}(z_\tau^m),$$

$z_\mu^k = x_\mu^k + iy_\mu^k$, $z_\tau^m = x_\tau^m + iy_\tau^m$ are complex coordinates of the points P_μ^k and P_τ^m (see Sec. 4).

4. In accordance with (5.1), (8.17) we write out, on every closed basic block $\overline{T}_\mu^*$, $1 \leq \mu \leq M$, the approximate solutions $u_{\mu n}^0$, $u_{\mu n}^\lambda$, $\lambda = 2, 3, \ldots, t$, of the corresponding problems (13.7), (12.1):

$$u_{\mu n}^0 = Q_{\mu 0}(z) + \beta_\mu \sum_{k=1}^{n_\mu} (u_\mu^{k0} - Q_\mu^{k0}) R_\mu(r_\mu, \theta_\mu, \theta_\mu^k), \qquad (13.13)$$

$$u_{\mu n}^\lambda = \sigma_\mu^\lambda + \beta_\mu \sum_{k=1}^{n_\mu} (u_\mu^{k\lambda} - Q_\mu^{k\lambda}) R_\mu(r_\mu, \theta_\mu, \theta_\mu^k), \qquad (13.14)$$

and the functions $v_{\mu n}^0$, $v_{\mu n}^\lambda$ conjugate to $u_{\mu n}^0$, $u_{\mu n}^\lambda$:

$$v_{\mu n}^0 = S_{\mu 0}(z) + \beta_\mu \sum_{k=1}^{n_\mu} (u_\mu^{k0} - Q_\mu^{k0}) I_\mu(r_\mu, \theta_\mu, \theta_\mu^k), \qquad (13.15)$$

$$v_{\mu n}^\lambda = \beta_\mu \sum_{k=1}^{n_\mu} (u_\mu^{k\lambda} - Q_\mu^{k\lambda}) I_\mu(r_\mu, \theta_\mu, \theta_\mu^k), \qquad (13.16)$$

where θ_μ^k are quantities (4.3), $\sigma_\mu^\lambda = 1$ for $\mu \in \Xi_\lambda$, $\sigma_\mu^\lambda = 0$ for $\mu \notin \Xi_\lambda$, $z = z(r_\mu, \theta_\mu) = x(r_\mu, \theta_\mu) + iy(r_\mu, \theta_\mu)$.

5. An approximate value ϑ_j^λ of the increment on the side γ_j, $N_1 + 1 \leq j \leq N$, of the function v^λ, conjugate to u^λ, $\lambda = 0, 2, 3, \ldots, t$,

can be found as follows. The vertices P_j and P_{j+1}, which are simultaneously the vertices of the basic block-sectors T_j^* and T_{j+1}^*, serve as the endpoints of the side γ_j of the polygon Ω. By condition IV (Sec. 2) imposed on the covering of the polygon Ω either $\overline{T}_j^* \cap \overline{T}_{j+1}^* \neq \varnothing$, or $\overline{T}_j^* \cap \overline{T}_{j+1}^* = \varnothing$. But then there is a finite number of block-half-disks $\overline{T}_\mu^*$ with some indices $\mu = a_j, b_j, \ldots, c_j, d_j$, each with its diameter on the side γ_j, and $\overline{T}_j^* \cap \overline{T}_{a_j}^* \neq \varnothing$, $\overline{T}_{a_j}^* \cap \overline{T}_{b_j}^* \neq \varnothing$, $\ldots$, $\overline{T}_{d_j}^* \cap \overline{T}_{j+1}^* \neq \varnothing$. In the first case we choose an arbitrary point $P_j' \in \overline{T}_j^* \cap \overline{T}_{j+1}^* \cap \gamma_j$ and set

$$\vartheta_j^\lambda = (v_{jn}^\lambda(P_j') - v_{jn}^\lambda(P_j)) + (v_{j+1,n}^\lambda(P_{j+1}) - v_{j+1,n}^\lambda(P_j')).$$

In the second case we specify some points

$$A_j \in \overline{T}_j^* \cap \overline{T}_{a_j}^* \cap \gamma_j, \qquad B_j \in \overline{T}_{a_j}^* \cap \overline{T}_{b_j}^* \cap \gamma_j, \qquad \ldots,$$

$$D_j \in \overline{T}_{c_j}^* \cap \overline{T}_{d_j}^* \cap \gamma_j, \qquad E_j \in \overline{T}_{d_j}^* \cap \overline{T}_{j+1}^* \cap \gamma_j$$

and set

$$\begin{aligned}\vartheta_j^\lambda = {}& (v_{jn}^\lambda(A_j) - v_{jn}^\lambda(P_j)) + (v_{a_jn}^\lambda(B_j) - v_{a_jn}^\lambda(A_j)) \\ &+ \ldots + (v_{d_jn}^\lambda(E_j) - v_{d_jn}^\lambda(D_j)) + (v_{j+1,n}^\lambda(P_{j+1}) - v_{j+1,n}^\lambda(E_j)).\end{aligned}$$

6. The approximate value $\tilde{\varkappa}_l^\lambda$ of the increment of the function v^λ conjugate to u^λ is computed under the traverse of the closed polygonal line Γ_l:

$$\tilde{\varkappa}_l^\lambda = \sum_{j=N_{l-1}+1}^{N_l} \vartheta_j^\lambda, \qquad 2 \leq l \leq t, \qquad \lambda = 0, 2, 3, \ldots, t. \qquad (13.17)$$

On the basis of Lemma 8.3, which can be directly extended to the boundary-value problems (12.1) and remains valid in the case of problem (13.7) for the covering of the polygon Ω specified in item 1, we have

$$|\tilde{\varkappa}_l^\lambda - \varkappa_l^\lambda| \leq \tilde{c} \exp\{-d_0 n\}, \qquad (13.18)$$

where $2 \leq l \leq t$, $\lambda = 0, 2, 3, \ldots, t$, $\varkappa_l^\lambda$ is quantity (12.2) (see Remark 12.1), $\tilde{c}, d_0 > 0$ are constants independent of the choice of the points P_j' or $A_j, B_j, \ldots, D_j, E_j$ and of n, for which system (13.12) is uniquely solvable.

7. We seek a solution $\tilde{c}_2'$, $\tilde{c}_3'$, ..., $\tilde{c}_t'$ of the system

$$\sum_{\lambda=2}^{t} \tilde{\varkappa}_l^\lambda \tilde{c}_\lambda' = -\tilde{\varkappa}_l^0, \qquad l = 2, 3, \ldots, t. \tag{13.19}$$

By virtue of Lemmas 4.1 and 12.2 and inequality (13.18) there exists an $\tilde{n}_0$ such that there is a unique solution of system (13.12) and (13.19) for any $n \geq \tilde{n}_0$ and

$$\max_{2 \leq \lambda \leq t} |\tilde{c}_\lambda' - c_\lambda'| \leq c' \exp\{-d_0 n\}, \tag{13.20}$$

where c_λ', $\lambda = 2, 3, \ldots, t$, is the solution of system (13.10), c', $d_0 > 0$ do not depend on n.

8. On every closed basic block $\overline{T}_\mu^*$, $1 \leq \mu \leq M$, we define a function

$$u_{\mu n} = \tilde{c}_1 + u_{\mu n}^0 + \sum_{\lambda=2}^{t} \tilde{c}_\lambda' u_{\mu n}^\lambda, \tag{13.21}$$

where $u_{\mu n}^0$, $u_{\mu n}^\lambda$ are functions (13.13), (13.14),

$$\tilde{c}_1 = -u_{\mu_0 n}^0(z_0) - \sum_{\lambda=2}^{t} \tilde{c}_\lambda' u_{\mu_0 n}^\lambda(z_0), \tag{13.22}$$

and μ_0 is the index of some closed basic block-disk $\overline{T}_{\mu_0}^*$ containing the point z_0.

9. On every block $\overline{T}_\mu^*$, $1 \leq \mu \leq M$, we define a function

$$v_{\mu n} = C_{\mu n} + v_{\mu n}^0 + \sum_{\lambda=2}^{t} \tilde{c}_\lambda' v_{\mu n}^\lambda, \tag{13.23}$$

conjugate to $u_{\mu n}$, where $v_{\mu n}^0$, $v_{\mu n}^\lambda$ are functions (13.15), (13.16), and $C_{\mu n}$ is an arbitrary constant.

10. We piece functions (13.23) together as follows. The constant $C_{\mu_0 n}$ is found from the requirement that $v_{\mu_0 n}(z_0) = 0$ (μ_0 is the same as in item 8). We consider all blocks $\overline{T}_\mu^*$ such that $\overline{T}_\mu^* \cap \overline{T}_{\mu_0}^* \cap \Omega \neq \varnothing$, and choose constants $C_{\mu n}$ from the condition of coincidence of the values of the functions $v_{\mu n}$ and $v_{\mu_0 n}$ at an arbitrary point of the set $\overline{T}_\mu^* \cap \overline{T}_{\mu_0}^*$. Then we take all blocks $\overline{T}_m^*$, which intersect in Ω with at least one of the blocks $\overline{T}_\mu^*$, for which the constant $C_{\mu n}$ is defined, and find the corresponding constants C_{mn} from the condition

of coincidence of v_{mn} with one of the completely defined functions $v_{\mu n}$ at an arbitrary point of the set $\overline{T}_m^* \cap \overline{T}_\mu^*$ ($\overline{T}_m^* \cap \overline{T}_\mu^* \cap \Omega \neq \varnothing$) and so on. By virtue of condition IV (Sec. 2) imposed on the covering of the polygon Ω, all the constants $C_{\mu n}$, $1 \leq \mu \leq M$, are defined in some way.

11. On each block $\overline{T}_\mu^*$, $1 \leq \mu \leq M$, we define a function

$$\zeta_{\mu n} = 1/(z - z_0) + u_{\mu n} + i v_{\mu n}, \qquad (13.24)$$

where $u_{\mu n}$ is function (13.21), $v_{\mu n}$ is function (13.23), with an arbitrary constant chosen in item 10.

Theorem 13.1. *For $n \geq \tilde{n}_0$ ($\tilde{n}_0$ is indicated in item 7)*

$$|\zeta_{\mu n} - \zeta| \leq c \exp\{-d_0 n\} \quad on \quad \overline{T}_\mu^*, \qquad (13.25)$$

where $1 \leq \mu \leq M$, $\zeta_{\mu n}$ is function (13.24), ζ is function (13.1), defining the desired conformal mapping, c, $d_0 > 0$ are constants independent of n.

Theorem 13.1 follows from Theorem 5.1 and Lemma 8.3 which can be directly extended to the boundary-value problems (12.1) and remaining valid for problem (13.7) for the covering of the polygon Ω specified in item 1, with due account of (13.2), (13.4), (13.8), (13.11)–(13.16), (13.20)–(13.23).

Remark 13.1. In the case of a simply-connected polygon Ω ($t = 1$) the operations described in items 5–7 are omitted and the sums in (13.8), (13.11), (13.21)–(13.23) are absent.

14. Approximate Conformal Mapping of a Multiply-Connected Polygon onto a Ring with Cuts along the Arcs of Concentric Circles

14.1. Structure of a Mapping

Suppose a t-connected polygon Ω is given in a complex plane z, $t \geq 2$ (see Sec. 1, 12). There exists a conformal mapping of the polygon Ω onto a ring, lying in the complex plane ζ, with $t - 2$ cuts along the arcs of circles which have a center in common with the circles that form the ring (see [17, Ch. 5, Sec. 1], [44, Sec. 1], [21]). We can arbitrarily choose two closed polygonal lines that form part of the boundary of the polygon Ω, one of which passes into the external boundary of the ring and the other into the internal boundary. We assume, for definiteness, that the external boundary of the polygon Ω formed by the polygonal line Γ_1 is associated with the external boundary of the ring and a part of the boundary formed by the polygonal line Γ_p, $p \neq 1$, is associated with the internal boundary of the ring. Without loss of generality we assume that the external boundary of the ring coincides with the circle $|\zeta| = 1$, and the vertex P_1 of the polygon Ω passes into the point $\zeta = 1$.

Let

$$\zeta = \zeta(z) \tag{14.1}$$

be the required mapping function. Since $\zeta(z) \neq 0$ on Ω, the function $\ln \zeta(z)$ has no branch points on the polygon Ω. We set

$$u = \ln |\zeta(z)|, \qquad v = \arg \zeta(z). \tag{14.2}$$

The function u is harmonic on Ω and assumes constant values on the polygonal lines Γ_l, $l = 1, 2, \ldots, t$, namely,

$$u = 0 \quad \text{on} \quad \Gamma_1, \qquad u = c_l \quad \text{on} \quad \Gamma_l, \qquad l = 2, 3, \ldots, t, \tag{14.3}$$

where the constants $c_l < 0$ are unknown.

The function v, conjugate to u, is multivalued on the polygon Ω, and

$$\delta_p v = -2\pi, \qquad \delta_l v = 0, \qquad l \neq 1, p, \tag{14.4}$$

where $\delta_q v$ is the increment of the function v under the traverse of the polygonal line Γ_q, when Ω remains locally on the left. Moreover, the condition

$$v(P_1) = 0 \tag{14.5}$$

is fulfilled for some branch of the function v.

The function u can evidently be represented in the form

$$u = \sum_{\lambda=2}^{t} c_\lambda u^\lambda, \tag{14.6}$$

where u^λ is the solution of the boundary-value problem (12.1). According to (14.4) the constants c_λ satisfy the equations

$$\sum_{\lambda=2}^{t} c_\lambda \varkappa_l^\lambda = -2\pi\delta_p^l, \qquad l = 2,3,\ldots,t, \qquad (14.7)$$

where $\varkappa_l^\lambda$ is quantity (12.2) (see Remark 12.1), δ_p^l is Kronecker's delta. By virtue of Lemma 12.2 the constants c_λ, $\lambda = 2,3,\ldots,t$, appearing, in particular, in (14.3), are uniquely defined by system (14.7). By virtue of (12.4), (12.5) the inequalities

$$c_p < c_\lambda < 0, \qquad \lambda = 2,3,\ldots,p-1,p+1,\ldots,t, \qquad (14.8)$$

hold true.

Therefore the desired function u is unique under our assumptions. Its conjugate function v is not single-valued. According to (14.4) the increment of v is equal to 2π when any simple contour lying in Ω and containing the polygonal line Γ_p in its interior is traversed counterclockwise. If the contour does not contain the polygonal line Γ_p in its interior, then the increment v is zero. When requirement (14.5) is met for some branch of the function v, this function is completely defined.

Thus, the required conformal mapping is defined by the function

$$\zeta = \exp\{u + iv\} \qquad (14.9)$$

and is unique. Note that the radius of the circle that forms the inner boundary of the ring (onto which the closed polygonal line Γ_p is mapped) is equal to

$$\rho_p = |\zeta|\big|_{\Gamma_p} = \exp\{c_p\}. \qquad (14.10)$$

Respectively, the radius of the circle on which the (two-fold) arc-image of the closed polygonal line Γ_λ lies, $\lambda \neq 1, p$, is

$$\rho_\lambda = |\zeta|\big|_{\Gamma_\lambda} = \exp\{c_\lambda\}, \qquad (14.11)$$

and, according to (14.8),

$$\rho_p < \rho_\lambda < 1, \qquad \lambda \neq 1, p.$$

14.2.   Constructing an Approximate Mapping

Here is the exposition of the approximate method of finding function (14.9) on blocks which has much in common with the corresponding approximate method from Sec. 13.

1. Without additional restrictions, the methods given in Sec. 2 are used to construct a covering of the polygon Ω.

2. A solution of system (13.12) is sought for each $\lambda = 2, 3, \ldots, t$.

3. Functions (13.14), (13.16) are constructed for $\mu = 1, 2, \ldots, M$; $\lambda = 2, 3, \ldots, t$.

4. The technique shown in items 5 and 6 of Sec. 13 is used to calculate quantities (13.17) for $\lambda, l = 2, 3, \ldots, t$.

5. The solution $\tilde{c}_2, \tilde{c}_3, \ldots, \tilde{c}_t$ is sought for the system

$$\sum_{\lambda=2}^{t} \tilde{\varkappa}_l^\lambda \tilde{c}_\lambda = -2\pi \delta_p^l, \qquad l = 2, 3, \ldots, t, \qquad (14.12)$$

which approximate system (14.7).

By virtue of Lemmas 4.1, 8.3, and 12.2 there is an $\tilde{n}_0$, such that system (13.12) has a unique solution for $\lambda = 2, 3, \ldots, t$ when $n \geq \tilde{n}_0$; in addition, inequalities (13.18) are satisfied for $l = 2, 3, \ldots, t$ for the indicated λ; finally, system (14.12) is uniquely solvable, and we have an estimate

$$\max_{2 \leq \lambda \leq t} |\tilde{c}_\lambda - c_\lambda| \leq c' \exp\{-d_0 n\}, \qquad (14.13)$$

where c_λ, $\lambda = 2, 3, \ldots, t$, is the solution of system (14.7), c', $d_0 > 0$ do not depend on n.

6. Two functions

$$u_{\mu n} = \sum_{\lambda=2}^{t} \tilde{c}_\lambda u_{\mu n}^\lambda, \qquad (14.14)$$

$$v_{\mu n} = C_{\mu n} + \sum_{\lambda=2}^{t} \tilde{c}_\lambda v_{\mu n}^\lambda, \qquad (14.15)$$

where $u_{\mu n}^\lambda$, $v_{\mu n}^\lambda$ are functions (13.14), (13.16), $C_{\mu n}$ is an arbitrary constant, are defined on every closed basic block $\overline{T}_\mu^*$, $1 \leq \mu \leq M$.

7. Functions (14.15) are pieced together in the way shown in item 10 of Sec. 13. Here we formally set $\mu_0 = 1$ and take the complex coordinate of the vertex P_1 of the polygon Ω to be z_0.

8. The function

$$\zeta_{\mu n} = \exp\{u_{\mu n} + i v_{\mu n}\}, \qquad (14.16)$$

where $u_{\mu n}$ is function (14.14), $v_{\mu n}$ is function (14.15), for which the arbitrary constant was chosen in item 7, is defined on every block $\overline{T}_\mu^*$, $1 \le \mu \le M$.

Theorem 14.1. *For* $n \ge \tilde{n}_0$ *($\tilde{n}_0$ was indicated in item 5) we have an inequality*

$$|\zeta_{\mu n} - \zeta| \le c \exp\{-d_0 n\} \quad on \quad \overline{T}_\mu^*, \qquad (14.17)$$

where $1 \le \mu \le M$, $\zeta_{\mu n}$ *is function (14.16),* ζ *is function (14.9) which defines the required conformal mapping,* c, $d_0 > 0$ *are constants which do not depend on* n.

Theorem 14.1 follows from (14.6), (14.13)–(14.15), Theorem 5.1 and Lemma 8.3.

Let $n \ge \tilde{n}_0$,

$$\tilde{\rho}_\lambda = \exp\{\tilde{c}_\lambda\}, \qquad \lambda = 2, 3, \ldots, t. \qquad (14.18)$$

Then by virtue of (14.10), (14.11), (14.13) we have an inequality

$$\max_{2 \le \lambda \le t} |\tilde{\rho}_\lambda - \rho_\lambda| \le c' \exp\{-d_0 n\}, \qquad (14.19)$$

where c', $d_0 > 0$ do not depend on n.

Remark 14.1. The conformal invariant of the doubly-connected polygon Ω ($t = 2$), i.e., the ratio ρ_1/ρ_2, where ρ_1, ρ_2 are the radii of the circles that form the external and internal boundaries of the ring onto which this polygon can be mapped, is approximately equal to

$$1/\tilde{\rho}_2 = \exp\{-\tilde{c}_2\} = \exp\{2\pi/\tilde{\varkappa}_2^2\}$$

($\rho_1 = 1$ in this case).

By virtue of (14.10), (14.13), (14.18) the inequality

$$|1/\tilde{\rho}_2 - 1/\rho_2| \le c'' \exp\{-d_0 n\},$$

where c'', $d_0 > 0$ do not depend on n, is satisfied for $n \ge \tilde{n}_0$.

Chapter 3

Development and Application of the Approximate Block Method for Conformal Mapping of Simply-Connected and Doubly-Connected Domains

15.   Approximate Conformal Mapping of Some Polygons onto a Strip

15.1.   Scheme of Constructing a Mapping

Suppose Ω is an open simply-connected polygon which has a simple boundary $\sigma = \sigma_0 \cup \sigma_1$, where σ_0, σ_1 are some of its connected parts, and set $\sigma_0 \cap \sigma_1$ consists of two points, u is a bounded solution of the boundary-value problem

$$\Delta u = 0 \quad \text{on} \quad \Omega$$

$$u = -\pi/4 \quad \text{on} \quad \sigma \setminus \sigma_1, \qquad u = \pi/4 \quad \text{on} \quad \sigma \setminus \sigma_0,$$

$$(15.1)$$

v is a (single-valued) harmonic function conjugate to u.

Then, evidently, the function

$$\zeta = \zeta(z) = u + i(v + C),$$

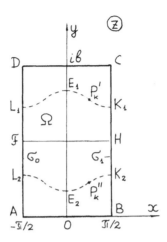

Figure 26

where C is an arbitrary constant, conformally maps the polygon Ω lying in the complex plane z, $z = x + iy$, onto the strip $|\mathrm{Re}\,\zeta| < \pi/4$. Under the mapping, the sraight line $\mathrm{Re}\,\zeta = (-1)^k \pi/4$, $k = 0, 1$, is the image of the set $\sigma \setminus \sigma_k$. The common endpoints of the polygonal lines σ_0, σ_1 go to infinity. We can (uniquely) choose the constant C such that the point defined on $\sigma \setminus \sigma_0$ is mapped into the point $\zeta = \pi/4$.

15.2.   Mapping of a Rectangle

Figure 26 shows a rectangle Ω with base $a = \pi$ and height $b \geq a$. Let σ_0 and σ_1 be sets of boundary points of the rectangle Ω lying on the imaginary axis and, respectively, on the left and the right of it.

To find an approximate solution of problem (15.1), we proceed as follows. As an auxiliary stage, we conformally map the rectangle Ω onto the half-ellipse Ω_0 (see Fig. 27) by means of the function

$$w = w(z) = (\sinh b)^{-1/2} \sin z. \tag{15.2}$$

In other words, we make a change of an independent variable by formula (15.2).

Instead of a solution of the boundary-value problem (15.1) we seek (approximately) the harmonic function, bounded on Ω_0, whose

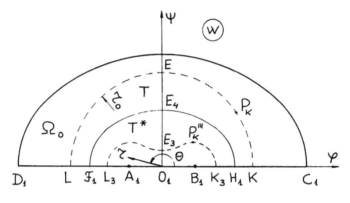

Figure 27

value on the part of the boundary of the half-ellipse Ω_0 to the left of the imaginary axis is $-\pi/4$ and on the part of the boundary to the right of the imaginary axis is $\pi/4$. We retain the notation u for the required function since this function and the solution of the problem (15.1) assume the same values at the points related by (15.2).

In the half-ellipse Ω_0 we inscribe a block-half-disk (an extended block)

$$T = \{(r,\theta) : 0 < r < r_0, 0 < \theta < \pi\},$$

where $r = |w|$, $\theta = \arg w$, $r_0 = (\sinh b)^{-1/2} \sinh(5b/6)$. As the basic block T^* we take a half-ellipse $F_1 E_4 H_1$, whose inverse image, under the mapping (15.2), is the lower half $AFHB$ of the rectangle Ω. The basic block T^* is a part of the extended block T, separated by a positive distance from the curvilinear part of the boundary of the extended block (in accordance with Remark 6.4). As we shall see later, we do not have to introduce any other blocks. It is sufficient to find the harmonic function u and its conjugate v on the half-ellipse $F_1 E_4 H_1$ (Fig. 27) or, respectively, on the lower half $AFHB$ of the rectangle Ω (Fig. 26). On the upper half of the rectangle Ω we can define the functions u and v proceeding from the evenness of the function u relative to the midline FH of this rectangle.

We define the kernels $R_1(r,\theta,\eta)$, $I_1(r,\theta,\eta)$:

$$\frac{R_1}{I_1}(r,\theta,\eta) =$$

$$= \sum_{p=0}^{1}(-1)^p\left(\frac{R}{I}\left(\frac{r}{r_0},\theta,(-1)^p\eta\right) - \frac{R}{I}\left(\frac{r}{r_0},\theta,\pi-(-1)^p\eta\right)\right),$$

where R and I are the kernel of the Poisson integral (3.14) and its conjugate (8.13).

According to Theorem 3.1, with due regard for the oddness of the function u relative to the imaginary axis ψ, this function can be represented on the block-half-disk T as

$$u(r,\theta) = Q(\theta) + \int_{0}^{\pi/2}(u(r_0,\eta) - Q(\eta))R_1(r,\theta,\eta)\,d\eta, \qquad (15.3)$$

where

$$Q(\theta) = \pi/4 - \theta/2. \qquad (15.4)$$

We choose a natural number n (n is the principal parameter of the method) and define points P_k, $k = 1,2,\ldots,n$, with polar coordinates $r = r_k = r_0$, $\theta = \theta_k = (k-1/2)\beta$, arranged on the arc KE (see Fig. 27) with a constant angular step

$$\beta = \pi/2n. \qquad (15.5)$$

We denote by P_k' the inverse image of the point P_k, under mapping (15.2), that lies on the curve K_1E_1 (see Fig. 26). Suppose that P_k'' is a point symmetric with respect to the point P_k' about the horizontal midline FH of the rectangle Ω, and P_k''' is the image of the point P_k'' lying in the basic block T^* on the curve K_3E_3 (Fig. 27).

Since the solution of the boundary-value problem (15.1) is even relative to the midline FH of the rectangle Ω, the equality

$$u(P_k) = u(P_k'''), \qquad k = 1,2,\ldots,n, \qquad (15.6)$$

holds true.

Let r_k''', θ_k''' be the polar coordinates of the point P_k''' and let u^k be the required approximate value of the function u (in accordance with (15.6)) at two points P_k and P_k''', $k = 1,2,\ldots,n$.

We set $r = r_k'''$, $\theta = \theta_k'''$, in (15.3) and (approximately) replace the integral in the resulting equality by the quardrature formula of rectangles with nodes P_m (which have polar coordinates $r = r_m = r_0$, $\theta = \theta_m = (m-1/2)\beta$), $m = 1, 2, \ldots, n$. Then we replace the unknown values $u(P_k''')$, $u(P_m)$ of the function u by their required approximate values u^k and, respectively, u^m, retaining the equality sign. Repeating this procedure for $k = 1, 2, \ldots, n$, we arrive at the following system of linear algebraic equations for the approximate values u^k of the function u at the points P_k:

$$u^k = Q(\theta_k''') + \beta \sum_{m=1}^{n} (u^m - Q(\theta_m))R^{km}, \qquad 1 \le k \le n, \qquad (15.7)$$

where Q is function (15.4), β is angular step (15.5), $\theta_m = (m-1/2)\beta$,

$$R^{km} = R_1(r_k''', \theta_k''', \theta_m). \qquad (15.8)$$

Thus the existence of equality (15.6), that follows from the evenness of the solution of the boundary-value problem (15.1) on the rectangle Ω relative to its midline FH, made it possible to express the approximate values u^k of the function u at the points P_k, $k = 1, 2, \ldots, n$, in terms of themselves and in this way set up system of equations (15.7).

In formula (15.8), corresponding to formula (4.7), the normalizing denominator is omitted since it is equal to unity in the cases considered below.

We seek a solution of system (15.7) by Seidel's method (see Sec. 4). We denote by u_ν^k, $k = 1, 2, \ldots, n$, the νth approximation of the solution of system (15.7) obtained by Seidel's method for an initial approximation $u_0^k = 0$, $k = 1, 2, \ldots, n$.

Let us consider a function

$$u_{n\nu}(r, \theta) = Q(\theta) + \beta \sum_{k=1}^{n} (u_\nu^k - Q(\theta_k))R_1(r, \theta, \theta_k), \qquad (15.9)$$

where $\theta_k = (k - 1/2)\beta$, harmonic on the half-disk T. This function differs from the approximate solution defined by formula (5.1) in that the exact solution of system (15.7) in its expression (15.9) is replaced by its approximation u_ν^k, $k = 1, 2, \ldots, n$, in Seidel's sense. Thus function (15.9) depends on the iteration number ν.

We substitute $r = |w(z)|$, $\theta = \arg w(z)$, where w is function (15.2), in (15.9) and get a function

$$u_{n\nu}(z) = \pi/4 - (\arg w(z))/2$$
$$+\beta \sum_{k=1}^{n}(u_\nu^k - Q(\theta_k))R_1(|w(z)|, \arg w(z), \theta_k) \qquad (15.10)$$

harmonic on the subdomain $ABK_1E_1L_1$ (see Fig. 26). On the intersection of the boundaries of the rectangle Ω and the indicated subdomain this function satisfies the boundary conditions defined in (15.1).

The harmonic function

$$v_{n\nu}(z) = (\ln|w(z)|)/2+$$
$$+\beta \sum_{k=1}^{n}(u_\nu^k - Q(\theta_k))I_1(|w(z)|, \arg w(z), \theta_k) \qquad (15.11)$$

is the conjugate of (15.10).

To follow the convergence of iterations when solving system (15.7), we calculate the control quantity

$$\varepsilon_{n\nu}^0 = \max_{1 \le p \le 21} |u_{n\nu}(z_p) - u_{n\nu}(z_p')|, \qquad (15.12)$$

where $z_p = \pi(2p-1)/84 + ib/3$, $z_p' = \pi(2p-1)/84 + i2b/3$ are points lying in the subdomain $ABK_1E_1L_1$ (Fig. 26) symmetrically about the midline FH of the rectangle Ω. At the points z_p and z_p' the values of the desired solution u obviously coincide.

In the process of solving system (15.7) the iterations terminate at the minimum $\nu \geq 2$ for which

$$\varepsilon_{n\nu}^0 \geq 0.98\varepsilon_{n,\nu-1}^0. \qquad (15.13)$$

There is no point in continuing iterations since the iteration process converges sufficiently rapidly and under condition (15.13) the principal part of $\varepsilon_{n\nu}^0$ is played by the intrinsic error of the block method rather than the inaccuracy in the solution of the algebraic problem (15.7). Henceforth ν will be the number determined by the method indicated above.

The value of $\varepsilon_{n\nu}^0$ just found is an indicator of the accuracy of the approximate solution $u_{n\nu}$. This value of $\varepsilon_{n\nu}^0$ is practically close to that of the quantity

$$E_n^0 = \max_{0 \le x \le \pi/2} |u_{n\nu}(x + ib/3) - u_{n\nu}(x + i2b/3)|,$$

in terms of which the following error estimates are expressed:

$$\max_{\substack{|x| \le \pi/2 \\ 0 \le y \le b/2}} |u_{n\nu}(x + iy) - u(x + iy)| \le \frac{3E_n^0}{2}, \qquad (15.14)$$

$$\max_{\substack{|x| \le \pi/2 \\ 0 \le y \le b/2}} |v_{n\nu}(x + iy) - v(x + iy) - c'| \le 10E_n^0, \qquad (15.15)$$

where u is the solution of the boundary-value problem (15.1) on the rectangle Ω being considered, v is its conjugate harmonic function, $u_{n\nu}$ and $v_{n\nu}$ are functions (15.10), (15.11), $c' = v_{n\nu}(ib/4) - v(ib/4)$. The values of the differences $u_{n\nu} - u$ and $v_{n\nu} - v$ at the point $z = 0$ are understood to be their limits at this point. For the sake of brevity, we omit the derivation of estimates (15.14), (15.15) which rests upon the principle of maximum, on estimates (29), (30) from [48] and on the properties of evenness and oddness of the function u. Note that the functions u and $u_{n\nu}$ are odd and v and $v_{n\nu}$ are even with respect to the imaginary axis.

Since $u_y'(x + ib/2) = 0$, $-\pi/2 \le x \le \pi/2$, we have $v(x + ib/2) =$ const . It is natural to verify how far the approximate conjugate function $v_{n\nu}$ is from the constant on the midline FH of the rectangle Ω (see Fig. 26). To this purpose, we calculate the second control quantity

$$\varepsilon_{n\nu}^1 = \max_{1 \le q \le 20} |v_{n\nu}(z_q) - v_{n\nu}^0|, \qquad (15.16)$$

where $z_q = \pi(2q - 1)/80 + ib/2$,

$$v_{n\nu}^0 = \frac{1}{20} \sum_{q=1}^{20} v_{n\nu}(z_q) \qquad (15.17)$$

taking into account the evenness of the function $v_{n\nu}$ relative to the imaginary axis.

The approximate mapping of the lower half of the rectangle Ω, i.e., the quadrilateral $AFHB$ (see Fig. 26), onto the half-strip $|\text{Re } \zeta| < \pi/4$, $\text{Im } \zeta < 0$ is defined by the function

$$\zeta = \zeta_n(z) = u_{n\nu}(z) + i(v_{n\nu}(z) - v_{n\nu}^0), \qquad (15.18)$$

where $u_{n\nu}$, $v_{n\nu}$ are functions (15.10), (15.11), and $v_{n\nu}^0$ is defined by (15.17). Under these conditions, (15.16) coincides with the maximum deviation of the images of forty points $z_q = \pi(2q - 1)/80 + ib/2$, $q = -19, -18, \ldots, 20$, lying on the midline FH (Fig. 26) from the real axis $\text{Im } \zeta = 0$. Under mapping (15.18), the images of the polygonal lines HBO and FAO (minus the point O) lie on the straight lines $\text{Re } \zeta = \pi/4$ and $\text{Re } \zeta = -\pi/4$, respectively, which form the boundary of the strip. The point O goes to infinity down the imaginary axis.

The approximate conformal mapping of the upper half of the rectangle Ω (the quadrilateral $FHCD$) onto the half-strip $|\text{Re } \zeta| < \pi/4$, $\text{Im } \zeta > 0$ is performed with the aid of the function

$$\zeta = \overline{\zeta_n(\bar{z} + ib)},$$

where ζ_n is function (15.18). This completes the construction of the approximate conformal mapping of the given rectangle Ω (Fig. 26) onto the strip $|\text{Re } \zeta| < \pi/4$.

The main results of the realization of the described algorithm are presented in Table 15.1. The first column gives the ratio of the sides of the rectangle Ω being mapped. Subsequent columns give the number of points n, the number of iterations ν needed to solve system (15.7), the control quantities $\varepsilon_{n\nu}^0$, $\varepsilon_{n\nu}^1$, and, finally, the quantity

$$h_n = \max_{1 \leq k \leq n} |u_\nu^k - Q(\theta_k)|,$$

equal to the maximum modulus of the coefficients under the summation sign in expressions (15.10) and (15.11) of the functions $u_{n\nu}$, $v_{n\nu}$.

Table 15.1

b/a	n	ν	$\varepsilon^0_{n\nu}$	$\varepsilon^1_{n\nu}$	h_n
1	12	7	$9.1 \cdot 10^{-12}$	$1.2 \cdot 10^{-12}$	$1.8 \cdot 10^{-1}$
2	7	4	$5.5 \cdot 10^{-12}$	$1.5 \cdot 10^{-12}$	$6.2 \cdot 10^{-2}$
5	3	3	$1.9 \cdot 10^{-12}$	$1.4 \cdot 10^{-12}$	$2.7 \cdot 10^{-3}$
10	2	2	$2.1 \cdot 10^{-12}$	$3.7 \cdot 10^{-12}$	$1.1 \cdot 10^{-5}$
16	1	2	$1.9 \cdot 10^{-12}$	$5.5 \cdot 10^{-12}$	$2.7 \cdot 10^{-8}$

The table shows a rapid decrease in n and h_n with an increase in the ratio of the sides of the rectangle being mapped (with the control parameters $\varepsilon^0_{n\nu}$ and $\varepsilon^1_{n\nu}$ maintained at approximately the same level). Thus the sums in (15.10) and (15.11) act as corrections to the principal parts (the real and imaginary parts of the approximate conformal mapping (15.18)) defined by the functions

$$\pi/4 - (\arg w(z))/2, \qquad (\ln |w(z)|)/2 - v^0_{n\nu},$$

where $w(z)$ has form (15.2), and $v^0_{n\nu}$ is quantity (15.17).

Table 15.2 illustrates the behavior of the main indicators of accuracy of the approximate conformal mapping of the square ($b = a = \pi$) onto a strip for a number of successive values on n. It also shows the behavior of the quantity ρ_n equal to the distance between the image of the lower right vertex of the square and the point $(1 - i)2^{-1/2}$ under the additional mapping $t = \tan \zeta_n(z)$ (the mapping of the strip onto a unit disk). Table 15.2 illustrates the exponential convergence of the block method with respect to n. The algorithm was run on a BESM-6 computer. The computing time for each version was a

Table 15.2

n	ν	$\varepsilon_{n\nu}^0$	$\varepsilon_{n\nu}^1$	ρ_n
2	2	$1.2 \cdot 10^{-2}$	$1.5 \cdot 10^{-3}$	$8.9 \cdot 10^{-5}$
4	3	$1.7 \cdot 10^{-4}$	$8.2 \cdot 10^{-6}$	$5.7 \cdot 10^{-7}$
6	4	$3.5 \cdot 10^{-6}$	$7.4 \cdot 10^{-8}$	$5.3 \cdot 10^{-9}$
8	5	$5.7 \cdot 10^{-8}$	$7.8 \cdot 10^{-10}$	$5.8 \cdot 10^{-11}$
10	6	$7.3 \cdot 10^{-10}$	$9.3 \cdot 10^{-12}$	$9.1 \cdot 10^{-13}$

fraction of a second.

Paper [18] shows another method of approximate conformal mapping of a rectangle and other domains onto a disk, but the nature of convergence of this method has not been established. As distinct from the result given in Table 15.1, the results in [18] shows a drastic deterioration of accuracy with an increase in the ratio of sides of the rectangle being mapped (even when the number of unknowns is maintained at a level of over 35).

15.3. Mapping of Two Octagons onto a Strip and of a Γ-Shaped Domain onto a Half-Strip

Let us consider polygons shown in Fig. 28 A and 28 B which we denote by Ω. These polygons are composed of six unit squares whose boundaries are shown in dash lines. Arguments relevant only to the polygon in Fig. 28 A (Fig. 28 B) will be referred to as Case A (Case B, respectively).

Suppose that σ_0 is a polygonal line with endpoints O and F con-

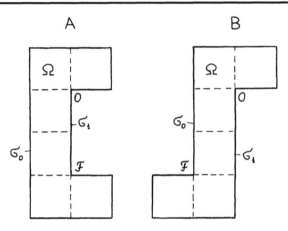

Figure 28

stituting a part of the boundary of the polygon Ω taken in the sense that the movement from the point O to F leaves Ω on the left, σ_1 is the remainder of the boundary of the polygon Ω, the points O and F inclusive.

To find an approximate solution of problem (15.1), we proceed as follows. We place the upper half of the polygon Ω in a single (extended) block-sector T of radius $r_0 = 1.75$ and angle $3\pi/2$ (see Fig. 29). We harmonically continue the solution u of problem (15.1) from Ω to the sector T, including the arc LK, retaining the same notation for the extended function. We have

$$u(x+iy) = \begin{cases} -\pi/2 - u(2 - x + iy) & \text{on} \quad A_0 LJ B_0, \\ -\pi/2 - u(x + i(2 - y)) & \text{on} \quad C_0 B_0 JCD, \\ u(x - i(2 + y)) & \text{on} \quad HKE_0 D_0 \quad \text{(Case A)}, \\ -u(-1 - x - i(2 + y)) & \text{on} \quad HKE_0 D_0 \quad \text{(Case B)}, \\ -\pi/2 - u(-2 - x + iy) & \text{on} \quad HDES. \end{cases}$$

$$(15.19)$$

This reflects the symmetry of the solution of problem (15.1) about the straight line $D_0 E_0$ in Case A and antisymmetry about the point M, lying midway between the points D_0 and E_0, in Case B. We accept the hexagon $D_0 E_0 O A_0 B_0 C_0$, constituting half of the given octagon Ω, as the basic block T^* (see Fig. 29). The basic block T^* is a part of the extended block-sector T.

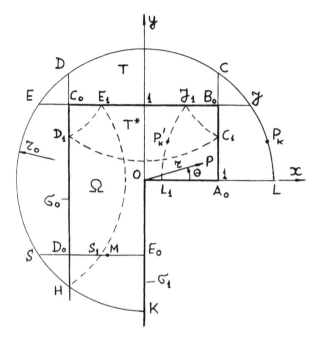

Figure 29

We define the kernels $R_2(r,\theta,\eta)$, $I_2(r,\theta,\eta)$:

$$\begin{matrix} R_2 \\ I_2 \end{matrix}(r,\theta,\eta) = \frac{2}{3}\sum_{p=0}^{1}(-1)^p \begin{matrix} R \\ I \end{matrix}\left(\left(\frac{r}{r_0}\right)^{2/3}, \frac{2\theta}{3}, (-1)^p\frac{2\eta}{3}\right),$$

where R, I have forms (3.14) and (8.13), respectively. These expressions for the kernels R_2, I_2 for the sector T are more simple than (3.17) and (8.16) since they refer to a special case of the boundary conditions of the first kind (on both sides of the sector).

By virtue of Theorem 3.1 the function u can be represented on the sector T as

$$u(r,\theta) = Q(\theta) + \int_0^{3\pi/2} (u(r_0,\eta) - Q(\eta))R_2(r,\theta,\eta)\,d\eta, \qquad (15.20)$$

where $r = |z|$, $\theta = \arg z$ ($z = x + iy$), $Q(\theta) = \theta/3 - \pi/4$. Note that the basic block $T^* \subset T$ (the hexagon $D_0E_0OA_0B_0C_0$) lies at a positive distance from the arc LK (Fig. 29) along which we carry out the integration in representation (15.20) (see Remark 6.4).

We choose a natural number n and specify points P_k, $k = 1, 2, \ldots, n$, with polar coordinates $r = r_k = r_0$, $\theta = \theta_k = (k - 1/2)\beta$ arranged on the arc LK with an angular step

$$\beta = 3\pi/2n. \tag{15.21}$$

With every point P_k belonging to the arc LJ (or the arc CD, or ES, respectively) we associate a point P'_k belonging to the arc $L_1 J_1$ (or $C_1 D_1$, or $E_1 S_1$, respectively) symmetric to it about the straight line $A_0 B_0$ (or the straight line $C_0 B_0$, or $D_0 C_0$, respectively). Here and in what follows we assume that the designated endpoints of each arc belong to that arc. Let any point P_k belonging to the arc JC (to the arc DE) be associated with a point P'_k, which belongs to the arc $J_1 C_1$ (to $D_1 E_1$) and is symmetric to it about the point B_0 (or the point C_0, respectively). We set $P^0_k = P_k$ if $P_k \in HK$ (HK is an arc). If $P_k \in EK \setminus (ES \cup HK)$, then we take as P^0_k a point symmetric with respect to the point P_k about the straight line DH. Finally, if $P_k \in EK \setminus ES$, then, in Case A (in Case B), we take as P'_k a point symmetric with respect to the point P^0_k about the straight line $D_0 E_0$ (about the point M, respectively). Thus all points P'_k, $k = 1, 2, \ldots, n$, have got in the basic block T^* (the hexagon $D_0 E_0 O A_0 B_0 C_0$).

In accordance with (15.19), we have

$$u(P_k) = \begin{cases} -\pi/2 - u(P'_k), & P_k \in LJ \cup CD \cup EH, \\ u(P'_k), & P_k \in LK \setminus (LJ \cup CD \cup EH) \end{cases} \tag{15.22}$$

in Case A and

$$u(P_k) = \begin{cases} -\pi/2 - u(P'_k), & P_k \in LJ \cup CD \cup ES, \\ u(P'_k), & P_k \in LS \setminus (LJ \cup CD \cup ES), \\ -u(P'_k), & P_k \in HK, \\ -\pi/2 + u(P'_k), & P_k \in EK \setminus (ES \cup HK) \end{cases} \tag{15.23}$$

in Case B.

Suppose that r'_k, θ'_k are polar coordinates of the point P'_k, and u^k is the required approximate value of the function u at the point P_k, $k = 1, 2, \ldots, n$. Setting $r = r'_k$, $\theta = \theta'_k$ in (15.20) and approximating the integral by the quadrature formula of rectangles with nodes P_m, $m = 1, 2, \ldots, n$, with a subsequent substitution of u^m for the values $u(P_m)$ of the function u, we arrive at an approximate equality

$$u(P'_k) \approx Q(\theta'_k) + \beta \sum_{m=1}^{n} (u^m - Q(\theta_m))R^{km}, \qquad (15.24)$$

where $Q(\theta) = \theta/3 - \pi/4$, β is angular step (15.21), $\theta_m = (m - 1/2)\beta$, $R^{km} = R_2(r'_k, \theta'_k, \theta_m)$. For a fixed k, $1 \leq k \leq n$, we replace $u(P_k)$ on the left-hand side of equality (15.22) (Case A) or (15.23) (Case B) by its approximate value u^k and replace formally $u(P'_k)$ by its approximate expression (15.24) on the right-hand side. Combining the relations obtained in this way for $k = 1, 2, \ldots, n$, we arrive at the following system of linear algebraic equations for the approximate values u^k of the function u at the points P_k:

$$u^k = \varkappa_k + \delta_k \left(Q'_k + \beta \sum_{m=1}^{n} (u^m - Q_m)R^{km} \right), \qquad 1 \leq k \leq n, \quad (15.25)$$

where $Q'_k = \theta'_k/3 - \pi/4$, $Q_m = (m - 1/2)\beta/3 - \pi/4$, $\delta_k = 1$, if $u(P'_k)$ appears with the plus sign on the right-hand side of (15.22) (Case A) or on the right-hand side of (15.23) (Case B), $\delta_k = -1$ otherwise; $\varkappa_k$ is a quantity coincident with the free term on the right-hand side of the corresponding relation (15.22) or (15.23).

Suppose that u^k_ν, $k = 1, 2, \ldots, n$, is an approximate solution of system (15.25), obtained after the νth Seidel iteration ($u^k_0 = \delta_k/2 - \pi/4$, $k = 1, 2, \ldots, n$),

$$u_{n\nu}(r, \theta) = \frac{\theta}{3} - \frac{\pi}{4} + \beta \sum_{k=1}^{n} (u^k_\nu - Q_k)R_2(r, \theta, \theta_k),$$

$$v_{n\nu}(r, \theta) = -\frac{\ln r}{3} + \beta \sum_{k=1}^{n} (u^k_\nu - Q_k)I_2(r, \theta, \theta_k).$$

The control parameters $\varepsilon^0_{n\nu}$, $\varepsilon^1_{n\nu}$ are defined as follows. Points A_1, A_2, ..., A_{12} are marked off on the line segment A_0B_0 (see Fig. 29) at a fixed distance $1/11$ from one another and points A_{13}, A_{14}, ..., A_{36} are marked off on the polygonal line $B_0C_0D_0$ at a fixed distance $1/6$ from one another, the points A_{24} and A_{36} being coincident with the points

C_0 and D_0, respectively. In addition, points B_q, $q = 1, 2, \ldots, 24$, with abscissa

$$x = x_q = -\frac{1}{2} + \begin{cases} q/24, & 1 \le q \le 12, \\ (12 - q)/24, & 13 \le q \le 24 \end{cases}$$

are marked off on the line segment $D_0 E_0$.

In Case A we set

$$\varepsilon_{n\nu}^0 = \max_{1 \le p \le 36} |u_{n\nu}(A_p) + \pi/4|,$$

$$\varepsilon_{n\nu}^1 = \max\{|v_{n\nu}(M) - v_{n\nu}^0|, \max_{1 \le q \le 24} |v_{n\nu}(B_q) - v_{n\nu}^0|\},$$

where

$$v_{n\nu}^0 = \frac{1}{25}\left(v_{n\nu}(M) + \sum_{q=1}^{24} v_{n\nu}(B_q)\right).$$

The control parameter $\varepsilon_{n\nu}^0$ coincides with the maximum deviation of the function $u_{n\nu}$ at 36 points, lying on the polygonal line $A_0 B_0 C_0 D_0$, from the boundary condition $-\pi/4$ given in (15.1). The quantity $\varepsilon_{n\nu}^1$ is equal to the maximum deviation of the approximate conjugate function $v_{n\nu}$, at 25 points of the line segment $D_0 E_0$, from its mean value $v_{n\nu}^0$ at these points (in Case A, the function v conjugate to the solution u of problem (15.1) is constant on $D_0 E_0$).

In Case B

$$\varepsilon_{n\nu}^0 = \max\{|u_{n\nu}(M)|, \max_{1 \le p \le 36} |u_{n\nu}(A_p) + \pi/4|\},$$

$$\varepsilon_{n\nu}^1 = \max_{1 \le q \le 12} |(v_{n\nu}(B_q) + v_{n\nu}(B_{q+12}))/2 - v_{n\nu}(M)|.$$

Here $\varepsilon_{n\nu}^0$ also allows for the deviation of the approximate solution $u_{n\nu}$ from zero at the point M (Fig. 29) at which the solution u of problem (15.1) is zero by virtue of the central antisymmetry. For the same reason, the substitution of the function v which is the conjugate of u for $v_{n\nu}$ in the expression for $\varepsilon_{n\nu}^1$ yields zero.

The criterion according to which the iterative process of solving system (15.25) is terminated is

$$\varepsilon_{n\nu}^0 + \varepsilon_{n\nu}^1 \ge 0.98(\varepsilon_{n,\nu-1}^0 + \varepsilon_{n,\nu-1}^1), \qquad \nu \ge 2.$$

Let us denote by Ω' the upper half of the octagon Ω, i.e., the set of all points which belong to Ω and lie above the straight line $D_0 E_0$ (Fig. 29).

In Case A, an approximate conformal mapping of Ω' (a Γ-shaped domain) onto a half-strip $|\text{Re } \zeta| < \pi/4$, $\text{Im } \zeta > 0$ is defined by the function

$$\zeta = u_{n\nu} + i(v_{n\nu} - v_{n\nu}^0). \tag{15.26}$$

Under these conditions the quantity $\varepsilon_{n\nu}^0$ coincides with the maximum deviation of the images of the points A_p, $p = 1, 2, \ldots, 36$, from the straight line $\text{Re } \zeta = -\pi/4$, and $\varepsilon_{n\nu}^1$ is equal to the maximum deviation of the images of the points B_q, $q = 1, 2, \ldots, 24$, and M from the real axis $\text{Im } \zeta = 0$. The images of the line segments $A_0 O$ and $E_0 O$ (minus the point O) lie on the straight lines $\text{Re } \zeta = -\pi/4$ and $\text{Re } \zeta = \pi/4$, respectively (the point O goes to infinity upward along the imaginary axis).

In Case B, the conformal mapping of the subdomain Ω' onto a certain part of the strip $|\text{Re } \zeta| < \pi/4$ is approximately defined by the function

$$\zeta = u_{n\nu} + i(v_{n\nu} - v_{n\nu}(M)). \tag{15.27}$$

To obtain an approximate conformal mapping of the octagon Ω onto the strip $|\text{Re } \zeta| < \pi/4$ we complete the mapping of its lower part $\Omega \setminus \overline{\Omega}'$ by means of the same function (15.26) or (15.27) using the axial symmetry (about the straight line $D_0 E_0$) in Case A and its central symmetry (about the point M) in Case B.

Remark 15.1. The application of the analytical continuation of the solution of the boundary-value problem (15.1) from the octagons Ω and the use of the properties of its symmetry allowed us to get by with only one block-sector. The general method of constructing a covering for polygons presented in Sec. 2 will require a large number of blocks (see, for instance, the covering of an L-shaped domain shown in Fig. 22).

The practical results are presented below in Tables 15.3 (Case A) and Table 15.4 (Case B). The average computing time for each variant on the BESM-6 computer was approximately 3 s. The behavior of the control parameters (indicators of accuracy) $\varepsilon_{n\nu}^0$ and $\varepsilon_{n\nu}^1$ substantiates

Table 15.3

n	ν	$\varepsilon_{n\nu}^0$	$\varepsilon_{n\nu}^1$
20	5	$7.9 \cdot 10^{-3}$	$7.8 \cdot 10^{-3}$
40	7	$1.6 \cdot 10^{-5}$	$4.2 \cdot 10^{-5}$
60	7	$1.2 \cdot 10^{-7}$	$8.9 \cdot 10^{-8}$
80	9	$6.2 \cdot 10^{-9}$	$1.8 \cdot 10^{-9}$

Table 15.4

n	ν	$\varepsilon_{n\nu}^0$	$\varepsilon_{n\nu}^1$
20	4	$8.1 \cdot 10^{-3}$	$4.7 \cdot 10^{-3}$
40	7	$1.6 \cdot 10^{-5}$	$2.1 \cdot 10^{-5}$
60	7	$1.3 \cdot 10^{-7}$	$2.0 \cdot 10^{-8}$
80	9	$6.2 \cdot 10^{-9}$	$4.7 \cdot 10^{-10}$

the exponential convergence of the approximate conformal mapping by the parameter n.

In paper [18] a different method is used for constructing an approximate conformal mapping of a Γ-shaped domain onto a unit disk with an accuracy of 10^{-5}, the number of the required unknowns being 85.

16.   Scheme of Constructing a Conformal Mapping of a Doubly-Connected Domain onto a Ring

The scheme of constructing conformal mappings of finitely-connected polygons onto canonical domains by reducing them to the boundary-value problems for Laplace's equation is presented in Sec. 12–14. In this section the scheme is given briefly but in sufficient detail for the case of mapping a doubly-connected polygonal domain onto a ring.

Let Ω be a finite doubly-connected domain in the complex plane z, $z = x + iy$, bounded by simple closed piecewise-analytic curves Γ_0 (the external boundary) and Γ_1 (the internal boundary). It is known that the domain Ω can be conformally mapped onto a ring. However, the ratio of the radii of the circles that form the external and the internal boundary of the ring is not known in advance. This ratio uniquely depends on the domain Ω and is called a conformal invariant (modulus) of the domain (see [17, Ch. 5, Sec. 1, Theorems 1 and 2]).

We denote by $\zeta = \zeta(z)$ a function, analytic in Ω, which gives a conformal mapping of the domain Ω onto a ring and by M the conformal invariant of this domain. We assume that Γ_0 is mapped onto the external boundary of the ring formed by a circle of unit radius and Γ_1 is carried into the internal boundary of the ring, which is a circle of radius ρ_1, $0 < \rho_1 < 1$, and, consequently,

$$M = 1/\rho_1. \tag{16.1}$$

Since $\zeta(z) \neq 0$ on Ω, it follows that the function $\ln \zeta(z)$ has no branch points in Ω. We set

$$U = \ln |\zeta(z)|, \qquad V = \arg \zeta(z). \tag{16.2}$$

The function U is harmonic in Ω and, by virtue of Theorem 1 from [17, Ch. 5, Sec. 1] is continuous up to its boundary $\Gamma_0 \cup \Gamma_1$. Obviously, the function U assumes constant values on Γ_0 and Γ_1, namely,

$$U = 0 \quad \text{on} \quad \Gamma_0, \qquad U = \vartheta = \ln \rho_1 < 0 \quad \text{on} \quad \Gamma_1. \qquad (16.3)$$

The function V, which is the conjugate of U, is multi-valued on Ω but, by virtue of the theorem indicated above, with due account of the fact that $|\zeta(z)| \geq \rho_1 > 0$ on $\overline{\Omega}$, its every branch is continuous up to the boundary $\Gamma_0 \cup \Gamma_1$. Upon a clockwise traverse of any closed simple contour that lies in Ω and contains Γ_1 in its interior and also along Γ_1, the increment δV of the function V is

$$\delta V = -2\pi \qquad (16.4)$$

in accordance with (16.2).

Thus the mapping function can be represented as

$$\zeta = \exp\{U + iV\}. \qquad (16.5)$$

Function (16.5) can be constructed as follows. The solution u is found for the boundary value problem:

$$\Delta u = 0 \quad \text{on} \quad \Omega, \qquad u = 0 \quad \text{on} \quad \Gamma_0, \qquad u = 1 \quad \text{on} \quad \Gamma_1. \qquad (16.6)$$

The increment δv is determined for the harmonic function v conjugate to u upon a clockwise traverse of the contour Γ_1 (like V the function v is continuous up to Γ_1). In accordance with (16.3), (16.4) we set

$$U = \vartheta u, \qquad V = \vartheta v + c_0, \qquad (16.7)$$

where v is a fixed branch of the function conjugate to u,

$$\vartheta = -2\pi/\delta v, \qquad (16.8)$$

c_0 is a constant which is used to fix the angle of rotation of the ring. Obviously, the condition $\delta V = -2\pi$ is fulfilled.

According to (16.3), (16.1) we have

$$\rho_1 = \exp\{\vartheta\}, \qquad M = \exp\{-\vartheta\}. \qquad (16.9)$$

This scheme is approximately realized below when we map a number of doubly-connected domains onto a ring.

17. Mapping a Square Frame onto a Ring

In this section we present the algorithm of an approximate conformal mapping of a square frame (a square with a cut-out square) onto a ring. The use of the symmetry of a domain and of an analytic continuation of the required function makes it possible to construct an approximate mapping with the aid of only one block-sector. The approximate conformal mapping is expressed in terms of elementary functions.

Suppose that $z = x + iy$, $\Omega_a = \{z : |x| < a, |y| < a\}$, $\Omega = \Omega_2 \setminus \overline{\Omega}_1$ is a square frame, Γ_0 (Γ_1) is an external (internal) boundary of the domain Ω, u is the solution of the boundary-value problem (16.6) on Ω. The domain Ω and the solution u have four axes of symmetry. Therefore, it is sufficient to find the solution u and its conjugate harmonic function v on the trapezoid $KLFH$ which constitutes one-eighth of the domain Ω (see Fig. 30).

We place the trapezoid in the block-sector $AEJH$ of a radius $r_0 = 1.75$ (an extended block) which is partially outside of the domain Ω. We denote this sector by T. The solution u of problem (16.6) is analytically (harmonically) continued from Ω to T, the arc AJ inclusive, first as an odd function across the straight line BF and then as an odd function across the straight line L^*D. Obviously, the extended function u is even with respect to the straight line HE. As the basic block T^* we take the hexagon $KLFL^*K^*H$ which contains the trapezoid $KLFH$ indicated above and satisfies the conditions pointed out in Remark 6.4.

We choose a natural number n and, on the arc AE with an angular step

$$\beta = 3\pi/4n \qquad (17.1)$$

specify points P_k, $k = 1, 2, \ldots, n$, that have polar coordinates $r = r_k = r_0$, $\theta = \theta_k = (k - 1/2)\beta$ (r is the distance from the point H, θ is the angle reckoned from the ray HA). With every point P_k belonging to the arc AB (to the arc CD) we associate a point P'_k, which belongs to the arc A_1B_1 (on the arc C_1D_1) and is symmetric to it about the straight line KL (about the straight line LF). Let every point P_k lying on the arc BC (on the arc DE) be associated with a point P'_k

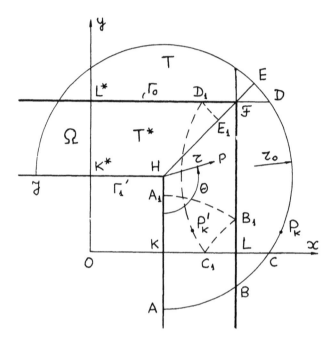

Figure 30

of the arc $B_1 C_1$ (of the arc $D_1 E_1$) which is symmetric to it about the
point L (about the point F). Obviously, all points P'_k, $k = 1, 2, \ldots, n$,
are in the closed basic block $\overline{T}^*$, and

$$u(P_k) = \begin{cases} -u(P'_k), & P_k \in BD, \\ u(P'_k), & P_k \in AE \setminus BD, \end{cases} \tag{17.2}$$

where AE, BD are arcs.

We define the kernels $R_1(r, \theta, \eta)$, $I_1(r, \theta, \eta)$:

$$\begin{aligned}
\frac{R_1}{I_1} (r, \theta, \eta) = \frac{2}{3} \sum_{p=0}^{1} (-1)^p \Big(\frac{R}{I} \Big(\Big(\frac{r}{r_0}\Big)^{2/3}, \frac{2\theta}{3}, \frac{(-1)^p 2\eta}{3} \Big) \\
+ \frac{R}{I} \Big(\Big(\frac{r}{r_0}\Big)^{2/3}, \frac{2\theta}{3}, \pi - \frac{(-1)^p 2\eta}{3} \Big) \Big),
\end{aligned}$$

where R, I are expressed by (3.14), (8.13).

By virtue of Theorem 3.1, with due regard for the evenness of the function u relative to the straight line HE (Fig. 30) this function can be represented as

$$u(r,\theta) = 1 + \int\limits_{0}^{3\pi/4} (u(r_0,\eta) - 1)R_1(r,\theta,\eta)\,d\eta \qquad (17.3)$$

on the sector T.

We denote the polar coordinates of the point P_k' by r_k', θ_k', and the desired approximate value of the function u at the point P_k, $k = 1, 2, \ldots, n$, by u^k. Setting $r = r_k'$, $\theta = \theta_k'$ in (17.3) and approximating the integral by the quadrature formula of rectangles with nodes P_m, $m = 1, 2, \ldots, n$, $\{P_m\}_{m=1}^{n} \equiv \{P_k\}_{k=1}^{n}$, with a subsequent substitution of u^m for $u(P_m)$, we obtain

$$u(P_k') \approx 1 + \beta \sum_{m=1}^{n} (u^m - 1)R^{km}, \qquad (17.4)$$

where β is angular step (17.1), $R^{km} = R_1(r_k', \theta_k', (m - 1/2)\beta)$.

The formal replacement of $u(P_k)$ by u^k in relation (17.2) and of $u(P_k')$ by its approximate expression (17.4) leads to a system of linear algebraic equations

$$u^k = \sigma_k(1 + \beta \sum_{m=1}^{n} (u^m - 1)R^{km}), \qquad 1 \le k \le n, \qquad (17.5)$$

where

$$\sigma_k = \begin{cases} -1, & P_k \in BD, \\ 1, & P_k \in AE \setminus BD, \end{cases}$$

relative to the approximate values of the function u at the points P_k.

Suppose that u_ν^k, $k = 1, 2, \ldots, n$, is the νth approximation of the solution of system (17.5) obtained by Seidel's method for $u_0^k = 2\sigma_k/5$,

$$u_{n\nu}(r,\theta) = 1 + \beta \sum_{k=1}^{n} (u_\nu^k - 1)R_1(r,\theta,\theta_k), \qquad (17.6)$$

$$v_{n\nu}(r,\theta) = \beta \sum_{k=1}^{n} (u_\nu^k - 1)I_1(r,\theta,\theta_k), \qquad (17.7)$$

where $\theta_k = (k - 1/2)\beta$.

Function (17.6) is an approximate solution of the boundary-value problem (16.6) on the closed basic block $\overline{T}^*$. The harmonic function (17.7) is the conjugate of function (17.6).

The solution u of problem (16.6) and its conjugate v (a single-valued branch) satisfy the following conditions on the sides of the trapezoid $KLFH$ (Fig. 30) belonging to T^*:

$$u = 1 \quad \text{on} \quad KH, \qquad u'_\theta = 0 \quad \text{on} \quad HF \setminus H, \qquad (17.8)$$

$$u = 0 \quad \text{on} \quad LF, \qquad (17.9)$$

$$v = c_1 \quad \text{on} \quad KL, \qquad (17.10)$$

$$v = c_2 \quad \text{on} \quad HF, \qquad (17.11)$$

where c_1, c_2 are constants.

It is easy to verify that for any natural ν the approximate solution (17.6) also satisfies conditions (17.8) and its conjugate function (17.7) satisfies condition (17.11) for $c_2 = 0$. As to conditions (17.9) and (17.10) for the functions $u_{n\nu}$ and $v_{n\nu}$, respectively, they are approximately satisfied.

The following control parameters are taken as indicators of the accuracy of the approximate solution $u_{n\nu}$ and its conjugate function $v_{n\nu}$:

$$\varepsilon_{n\nu}^0 = \max_{1 \le p \le 41} |u_{n\nu}(A_p)|, \qquad (17.12)$$

$$\varepsilon_{n\nu}^1 = \max_{1 \le q \le 31} |v_{n\nu}(B_q) - v_{n\nu}^0|, \qquad (17.13)$$

where

$$v_{n\nu}^0 = \frac{1}{15} \sum_{q=1}^{15} v_{n\nu}(B_q),$$

A_p, $p = 1, 2, \ldots, 41$, are points lying on the line segment LF with step $h = 1/20$, B_q, $q = 1, 2, \ldots, 31$, are points lying on the line segment KL with step $h_1 = 1/30$ and numbered from left to right, $v_{n\nu}^0$ is the mean value of the function $v_{n\nu}$ over the 15 chosen points on the left-hand side of KL.

Control quantities (17.12), (17.13) are sufficiently close to the maximum values of the modulus of the difference of the left-hand

and right-hand sides of (17.9), (17.10) in which the functions u and v are replaced by their approximations $u_{n\nu}$ and $v_{n\nu}$, with $c_1 = v^0_{n\nu}$.

The iteration of the solution of system (17.5) is terminated at the minimum $\nu \geq 2$ satisfying condition (15.13). This ν is fixed below.

We find the approximate values ϑ_n, ρ_{1n}, and M_n of quantities (16.8), (16.9) from the formulas

$$\vartheta_n = -\pi/4v^0_{n\nu}, \tag{17.14}$$

$$\rho_{1n} = \exp\{\vartheta_n\}, \qquad M_n = \exp\{-\vartheta_n\} \tag{17.15}$$

with due account of the fact that $v_{n\nu} \equiv 0$ on HF and, consequently, $v^0_{n\nu} - 0 \approx \delta v/8$.

The approximate conformal mapping of the basic block T^*, i.e., the hexagon $KLFL^*K^*H$, constituting a quarter of the domain Ω (see Fig. 30), onto a quarter of the ring,

$$\rho_{1n} < |\zeta| < 1, \qquad 0 < \arg\zeta < \pi/2 \tag{17.16}$$

is defined (in accordance with (16.5)) by the elementary analytic function

$$\zeta = \zeta_n(z) = \exp\{U_n + iV_n\}, \tag{17.17}$$

where

$$U_n = \vartheta_n u_{n\nu}, \qquad V_n = \pi/4 + \vartheta_n v_{n\nu}. \tag{17.18}$$

Function (17.17) carries points symmetric with respect to the straight line HF into points symmetric with respect to the ray $\arg\zeta = \pi/4$. It is obvious that the images of the points A_p, $p = 1, 2, \ldots, 41$, lying on the line segment LF deviate from the circle $|\zeta| = 1$ by not more than the quantity

$$E^0_n = \exp\{|\vartheta_n|\varepsilon^0_{n\nu}\} - 1, \tag{17.19}$$

and the arguments of the images of the points B_q, $q = 1, 2, \ldots, 31$, lying on the line segment KL deviate from zero by not more than

$$E^1_n = |\vartheta_n|\varepsilon^1_{n\nu}. \tag{17.20}$$

Table 17.1

n	ν	M_n	E_n^0	E_n^1
5	3	1.86	$7.0 \cdot 10^{-2}$	$9.4 \cdot 10^{-2}$
10	6	1.8476	$3.8 \cdot 10^{-3}$	$4.0 \cdot 10^{-3}$
30	9	1.847 709 02	$6.2 \cdot 10^{-8}$	$2.0 \cdot 10^{-8}$
50	11	1.847 709 011 235	$1.1 \cdot 10^{-10}$	$1.6 \cdot 10^{-11}$
M		1.847 709 011 236		
M^*		1.847 709 011 217		

The image of the polygonal line KHK^* lies entirely on the circle $|\zeta| = \rho_{1n}$.

The approximate conformal mapping of the hexagon $KLFL^*K^*H$ defined by function (17.17) extends to the whole domain Ω by means of the symmetry of this domain with respect to the real and imaginary axes.

Table 17.1 gives the results of calculations carried out according to the presented algorithm on BESM-6 computer in the single-precision mode (to approximately 12 or 13 decimal places).

Here n is the number of desired real parameters (the order of system (17.5)), ν is the number of iterations carried out in solving the system of linear algebraic equations (17.5), M_n is the approximate value of the conformal invariant of the domain Ω (the square frame),

E_n^0, E_n^1 are the indicators of accuracy of the approximate conformal mapping indicated above.

Table 17.1 also gives the value of the conformal invariant M of the domain Ω with 13 correct digits computed in [34] with the use of the corresponding formula from [8]. It also gives the approximate value M^* of the conformal invariant found in [34] by the orthogonalization method for the number of required complex parameters $n_{opt} = 24$.

Table. 17.1 illustrates the exponential convergence of the block method in the case when a domain has reentrant angles.

18.   Mapping a Square with a Circular Hole Using a Circular Lune Block

Let $0.2 \leq a \leq 0.6$, $z = x + iy$,

$$\Omega = \{z : |x| < 1, |y| < 1\} \setminus \{z : |z| \leq a\}. \tag{18.1}$$

We have $\Gamma_1 = \{z : |z| = a\}$, $\Gamma_0 = \overline{\Omega} \setminus (\Omega \cup \Gamma_1)$.

Since the domain Ω has four axes of symmetry, it is sufficient to find the solution u of the boundary-value problem (16.6) and its conjugate function v only on the circular quadrangle $KLFH$ (see Fig. 31) constituting one-eighth of this domain.

We place the circular quadrangle $KLFH$ in the right-angled circular lune $AEJH$ (an extended block) formed by the arc AHJ of the circle Γ_1 of radius a and the arc AEJ of the circle of radius

$$r_0 = \min \left\{ 0.95; \left(2 + \left(\frac{2a}{2 - a^2} \right)^2 \right)^{1/2} - \frac{2a}{2 - a^2} \right\}, \tag{18.2}$$

whose center O_1 lies on the ray OF.

We denote this circular lune block by T. We harmonically extended the solution u of problem (16.6) from Ω to the block T, including the arc AEJ, first as an odd function across the straight line SF and then as an odd function across the straight line L^*D. We take the circular pentagon $KLFL^*K^*$ as the basic block T^*.

We accept expression (18.2) for r_0 since the block T did not intersect the mirror reflection of the hole in (18.1) relative to the straight lines SF and L^*D, and for a small radius of the hole the block T is

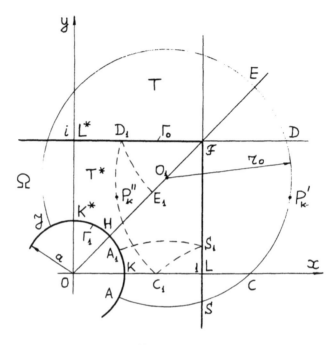

Figure 31

at a positive distance from these mirror reflections, near which the extended function u varies drastically in this case. In addition, we have $T^* \subset T$ for the chosen r_0, the basic block T^* being at a positive distance from the arc AEJ (see Remark 6.4).

Let

$$w = w(z) \tag{18.3}$$

be a fractional-linear function conformally mapping the block T onto the half-disk

$$|w| < 1, \qquad \operatorname{Im} w > 0 \tag{18.4}$$

so that the point A (Fig. 31) passes into the point $w = 1$ and the point H into the point $w = 0$. We denote the inverse function by $z(w)$.

Suppose furthermore that $u(z)$ is the solution of the boundary-value problem (16.6) on the domain Ω continued to the block T and

$\xi(w)$ is a function which is harmonic on the half-disk (18.4) and assumes the values

$$\xi = \begin{cases} 1, & \text{Im } w = 0, \\ u(z(w)), & |w| = 1, \end{cases} \tag{18.5}$$

on its boundary.

Obviously,

$$u(z(w)) \equiv \xi(w), \qquad \xi(w(z)) \equiv u(z), \qquad z \in \overline{T}. \tag{18.6}$$

We introduce a polar coordinate system

$$r = |w|, \qquad \theta = \arg w$$

on the half-disk (18.4).

We choose a natural number n and specify points P_k, $k = 1, 2, \ldots, n$, with coordinates $r = r_k = 1$, $\theta = \theta_k = (k - 1/2)\beta$, on the curvilinear part of the boundary of the half-disk (18.4) with angular step

$$\beta = \pi/2n. \tag{18.7}$$

Then we denote the inverse image of the point P_k (under mapping (18.3)), lying on the arc AE by P_k' (Fig. 31). With every point P_k' belonging to the arc AS (arc CD) we associate a point P_k'', which belongs to the arc $A_1 S_1$ (to $C_1 D_1$) and is symmetric to it about the straight line KL (the straight line LF). Let any point P_k' lying on the arc SC (on the arc DE) be associated with a point P_k'' belonging to the arc $S_1 C_1$ (to the arc $D_1 E_1$) and symmetric to it about the point L (or the point F, respectively). Note that the points P_k'', $k = 1, 2, \ldots, n$, lie in the closed basic block $\overline{T}^*$. Finally, we denote by P_k''', $k = 1, 2, \ldots, n$, the image of the point P_k'' under mapping (18.3) (belonging to the half-disk (18.4)).

By virtue of (18.6) and the obvious properties of evenness and oddness of the function u we have

$$\xi(P_k) = \begin{cases} -\xi(P_k'''), & P_k' \in SD, \\ \xi(P_k'''), & P_k' \in AE \setminus SD, \end{cases} \tag{18.8}$$

where AE, SD are arcs.

We define the kernels $R_1(r, \theta, \eta)$, $I_1(r, \theta, \eta)$:

$$\begin{array}{c} R_1 \\ I_1 \end{array} (r, \theta, \eta) = \sum_{p=0}^{1} (-1)^p \left(\begin{array}{c} R \\ I \end{array} (r, \theta, (-1)^p \eta) + \begin{array}{c} R \\ I \end{array} (r, \theta, \pi - (-1)^p \eta) \right),$$

where R, I are functions (3.14), (8.13).

By virtue of Theorem 3.1 with due regard for the evenness of the function ξ relative to the imaginary axis, this function can be represented on the half-disk (18.4) as

$$\xi(r, \theta) = 1 + \int_0^{\pi/2} (\xi(1, \eta) - 1) R_1(r, \theta, \eta) \, d\eta. \tag{18.9}$$

Approximating the integral in (18.9) by the quadrature formula of rectangles and taking into account (18.8) (see Sec. 17), we arrive at a system of linear algebraic equations for the approximate values ξ^k of the function ξ at the points P_k, or , what is the same, for the approximate values $u^k = \xi^k$ of the function u at the points P'_k. This system is similar to (17.5) and has the form

$$u^k = \sigma_k (1 + \beta \sum_{m=1}^{n} (u^m - 1) R^{km}), \qquad 1 \le k \le n, \tag{18.10}$$

where β is angular step (18.7), $R^{km} = R_1(r'''_k, \theta'''_k, (m - 1/2)\beta)$,

$$\sigma_k = \left\{ \begin{array}{ll} -1, & P'_k \in SD, \\ 1, & P'_k \in SE \setminus SD, \end{array} \right.$$

r'''_k, θ'''_k are the polar coordinates of the point P'''_k (in the complex plane w).

Suppose that u^k_ν, $k = 1, 2, \ldots, n$, is an approximate solution of system (18.10), obtained after the νth iteration in the sense of Seidel for the initial approximation $u^k_0 = \sigma_k/2$,

$$u_{n\nu}(z) = 1 + \beta \sum_{k=1}^{n} (u^k_\nu - 1) R_1(|w(z)|, \arg w(z), \theta_k), \tag{18.11}$$

$$v_{n\nu}(z) = \beta \sum_{k=1}^{n} (u^k_\nu - 1) I_1(|w(z)|, \arg w(z), \theta_k), \tag{18.12}$$

Table 18.1

a	n_{opt}	M
0.2	20	5.393 525 710 616
0.4	22	2.696 724 431 230
0.8	28	1.342 990 365 599

where β is angular step (18.7), $\theta_k = (k - 1/2)\beta$, w is function (18.3).

We consider function (18.11) to be an approximate solution of the boundary-value problem (16.6) on the closed basic block $\overline{T}^*$ (on the closed circular pentagon $KLFL^*K^*$). Function (18.12) is the conjugate of (18.11).

When solving system (18.14) by the iteration method, we calculate the control parameters

$$\varepsilon_{n\nu}^0 = \max_{1 \le p \le 31} |u_{n\nu}(z_p^0)|,$$

$$\varepsilon_{n\nu}^1 = \max_{1 \le q \le 31} |v_{n\nu}(z_q^1) - v_{n\nu}^0|,$$

where $z_p^0 = 1 + i(p - 1)/30$, $z_q^1 = a + (q - 1)(1 - a)/30$,

$$v_{n\nu}^0 = \frac{1}{16} \sum_{q=1}^{16} v_{n\nu}(z_q^1).$$

as we did in Sec. 17. The iterations are terminated at the minimum $\nu \ge 2$ satisfying condition (15.13).

The approximate values ϑ_n, ρ_{1n}, M_n of quantities (16.8), (16.9) can be found from formulas (17.14), (17.15). The approximate conformal mapping of the circular pentagon $KLFL^*K^*$, i.e., of a quarter of domain (18.1), on a quarter of ring (17.16) is defined by a function

Table 18.2

$(a = 0.4)$

n	ν	M_n	E_n^0	E_n^1
15	3	2.697	$8.9 \cdot 10^{-4}$	$1.3 \cdot 10^{-3}$
30	7	2.696 724 45	$1.8 \cdot 10^{-7}$	$7.9 \cdot 10^{-7}$
45	10	2.696 724 431 229	$5.3 \cdot 10^{-10}$	$1.5 \cdot 10^{-10}$

Table 18.3

$(a = 0.6)$

n	ν	M_n	E_n^0	E_n^1
10	4	1.797 3	$3.2 \cdot 10^{-2}$	$3.6 \cdot 10^{-2}$
30	5	1.797 176 9	$5.3 \cdot 10^{-5}$	$5.7 \cdot 10^{-5}$
50	7	1.797 177 548	$1.0 \cdot 10^{-7}$	$7.1 \cdot 10^{-8}$
70	10	1.797 177 558 085	$1.1 \cdot 10^{-10}$	$1.6 \cdot 10^{-10}$

of form (17.17). The accuracy indicators E_n^0, E_n^1 can be found from formulas (17.19), (17.20) and their sense is similar to that given in Sec. 17.

Before presenting the concrete results obtained by this method, we give in Table 18.1 the values of the conformal invariant M of domain (18.1) for three values of a found in [34] by the orthogonalization method and the variational method. For the two indicated methods these values coincide to within the 13 significant digits given. As was noted in [34], the variational method gives an upper approximation for the desired conformal invariant of the domain. The results in [34] were obtained on CDC 7600 computer. Table 18.1 also gives the number n_{opt} of the required complex-valued parameters (coefficients).

We shall compare some results obtained by the block method in this section, and also the results obtained in Sec. 20 and Sec. 21, with those given in Table 18.1.

Tables 18.2 and 18.3 present the main results of computations carried out according to the given algorithm on the BESM-6 computer in the single-precision mode. These tables illustrate the exponential convergence of the block method, its high accuracy and strong stability as concerns the computing rounding-off errors. The value M_{45} in Table 18.2 differs from the respective value M in Table 18.1 by a single unit in the least significant (13th) place. which lies within the limits of accuracy of the representation of numbers on the BESM-6 computer. The number ν of iterations carried out to solve system (18.10), which is defined by condition (15.13), is not large. Tables 18.2 and 18.3 give the number n of the required real parameters.

19.   Representation of a Harmonic Function on a Ring

Let $z = x + iy$, $r = |z|$, $\theta = \arg z$, $0 < l_1 < l_0$, and suppose that

$$K = \{(r,\theta) : l_1 < r < l_0, 0 \le \theta < 2\pi\} \qquad (19.1)$$

is a ring,

$$\sigma_j = \{(r,\theta) : r = l_j, 0 \le \theta < 2\pi\}, \qquad j = 0,1,$$

are the external and the internal boundary of the ring K.

Let us consider Dirichlet's problem

$$\Delta u = 0 \quad \text{on} \quad K, \tag{19.2}$$

$$u = \varphi_j(\theta) \quad \text{on} \quad \sigma_j, \quad j = 0, 1, \tag{19.3}$$

where $\varphi_j(\theta)$ are given continuous 2π-periodic functions.

According to Villat's formula (see [1]), the solution u of the boundary-value problem (19.2), (19.3) can be represented, on the ring K, in the form

$$u(r, \theta) = \frac{a_1 \ln l_0 - a_0 \ln l_1}{\ln l_0 - \ln l_1} + \frac{a_0 - a_1}{\ln l_0 - \ln l_1} \ln r$$

$$+ \sum_{j=0}^{1} (-1)^j \int_0^{2\pi} (\varphi_j(\eta) - a_j) \Re(r, \theta, l_j, h, \eta) \, d\eta, \tag{19.4}$$

where

$$a_j = \frac{1}{2\pi} \int_0^{2\pi} \varphi_j(\eta) \, d\eta, \tag{19.5}$$

$$h = \frac{l_1}{l_0}, \tag{19.6}$$

$$\Re(r, \theta, l, h, \eta) = \operatorname{Re} \zeta_0(t, h), \tag{19.7}$$

$$t = \frac{r}{l}(\cos(\theta - \eta) + i \sin(\theta - \eta)), \tag{19.8}$$

$$\zeta_0(t, h) = \frac{1}{2\pi} \left(\frac{1+t}{1-t} + 2 \sum_{k=1}^{\infty} \left(\frac{h^{2k}t}{1 - h^{2k}t} - \frac{h^{2k}t^{-1}}{1 - h^{2k}t^{-1}} \right) \right) \tag{19.9}$$

is Weierstrass' function.

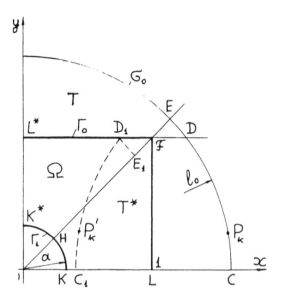

Figure 32

20.   Using a Block-Ring for Mapping Domain (18.1) onto a Ring

Suppose that the given a and chosen l_0 satisfy the inequalities

$$0 < a < 2 - \sqrt{2}, \qquad \sqrt{2} < l_0 \leq 2 - a. \qquad (20.1)$$

We set $l_1 = a$ and place domain (18.1) in ring (19.1) which we call an (extended) block and denote by T. The domain (18.1) itself, denoted also by Ω, plays the part of the basic block T^*. We extend the solution u of problem (16.6) on domain (18.1) from this domain analytically (harmonically) by the technique indicated in Sec. 18, to the block-ring T, including its boundary (this is possible under conditions (20.1)), and retain the same notation for the extended function.

On the arc CE of the circle σ_0 (see Fig. 32) we specify points P_k, $k = 1, 2, \ldots, n$, with polar coordinates $r = r_k = l_0$, $\theta = \theta_k = (k - 1/2)\beta$, where

$$\beta = \pi/4n \qquad (20.2)$$

is an angular step, n is a natural number. We associate the point P_k, lying on the arc CD, with a point P'_k of the arc C_1D_1, which is symmetric to it about the straight line LF. Let every point P_k belonging to the arc DE be associated with a point P'_k lying on the arc D_1E_1 and symmetric to it about the point F. Obviously, all points P'_k, $k = 1, 2, \ldots, n$, are on the closed domain $\overline{\Omega}$, i.e., on the closed basic block $\overline{T}^*$. Since the function u vanishes on the line segments LF and L^*D, we have

$$u(P_k) = \begin{cases} -u(P'_k), & P_k \in CD, \\ u(P'_k), & P_k \in CE \setminus CD, \end{cases} \tag{20.3}$$

where CE and CD are arcs.

Let $r = r'_k$, $\theta = \theta'_k$ be the polar coordinates of the point P'_k. We define the kernels $R_0(r, \theta, \eta)$, $I_0(r, \theta, \eta)$:

$$\begin{matrix} R_0 \\ I_0 \end{matrix} (r, \theta, \eta) = \sum_{\mu=1}^{8} \begin{matrix} \text{Re} \\ \text{Im} \end{matrix} \zeta_0\left(\frac{r}{l_0}t_\mu(\theta, \eta), h\right), \tag{20.4}$$

where ζ_0 is function (19.9), $h = a/l_0$,

$$t_\mu(\theta, \eta) = \exp\{i(\theta - \psi_\mu(\eta))\},$$

$$\psi_\mu(\eta) = \begin{cases} \eta + \pi(\mu - 1)/2, & \mu = 1, 2, 3, 4, \\ -\eta - \pi(\mu - 5)/2, & \mu = 5, 6, 7, 8. \end{cases}$$

Since the (extended) function u is even relative to the real and imaginary axes and the bisectors of the angles formed by these axes and since $u(a, \eta) \equiv 1$, $0 \le \eta < 2\pi$, the indicated function can by (19.4) be represented on the ring-block T as

$$u(r, \theta) = \frac{\ln l_0 - a_0 \ln a}{\ln l_0 - \ln a} + \frac{a_0 - 1}{\ln l_0 - \ln a} \ln r$$

$$+ \int_0^{\pi/4} (u(l_0, \eta) - a_0) R_0(r, \theta, \eta) \, d\eta, \tag{20.5}$$

where R_0 is the kernel defined in (20.4),

$$a_0 = \frac{4}{\pi} \int_0^{\pi/4} u(l_0, \eta) \, d\eta. \tag{20.6}$$

Setting $r = r'_k$, $\theta = \theta'_k$, $k = 1, 2, \ldots, n$, in (20.5) and replacing the integrals in (20.5), (20.6) by the quadrature formulas of rectangles with nodes $\eta = \theta_m = (m - 1/2)\beta$, $m = 1, 2, \ldots, n$, we arrive, by (20.3) (as we did in Sec. 17), at the following system of linear algebraic equations for the approximate values u^k of the function u at the points P_k:

$$u^k = \sigma_k \left(\frac{\ln l_0 - \ln r'_k}{\ln l_0 - \ln a} + \beta \sum_{m=1}^{n} u^m R^{km} \right), \qquad 1 \le k \le n, \qquad (20.7)$$

where β is angular step (20.2),

$$R^{km} = R_0^{km} - \frac{1}{n} \sum_{q=1}^{n} R_0^{kq} + \frac{4}{\pi} \frac{\ln r'_k - \ln a}{\ln l_0 - \ln a},$$

$$R_0^{km} = R_0(r'_k, \theta'_k, \theta_m),$$

$$\sigma_k = \left\{ \begin{array}{ll} -1, & P_k \in CD, \\ 1, & P_k \in CE \setminus CD. \end{array} \right.$$

Remark 20.1. By virtue of the symmetry properties of the function u mentioned above, the accepted approximation of the integrals in (20.5), (20.6) is equivalent to the approximation by the quadrature formulas of rectangles of the integrals in representations (19.4) and (19.5) corresponding to the periodic case, and in periodic cases the quadrature formula of rectangles is natural (see [24], and, in particular, Lemma 6.1).

Remark 20.2. By virtue of (20.1), the basic block $T^* \subset T$, coincident with the given domain (18.1), lies at a positive distance from the external boundary of the extended ring-block T, on which the points P_k, i.e., the nodes of the quadrature formula being used, lie. Thus the condition pointed out in Remark 6.4 is fulfilled.

Let u^k_ν, $k = 1, 2, \ldots, n$, be the approximate solution of system (20.7) obtained after the νth iteration in Seidel's sense for the zero initial approximation. The approximate solution $u_{n\nu}$ of problem (16.6) on domain (18.1) and its conjugate harmonic function $v_{n\nu}$ have the form

Table 20.1

a	l_0	n	ν	$M - M_n$	E_n^0
0.2	1.7	18	13	$1 \cdot 10^{-12}$	$2.2 \cdot 10^{-11}$
0.4	1.6	23	11	$1 \cdot 10^{-12}$	$1.7 \cdot 10^{-10}$

$$u_{n\nu}(r,\theta) = \frac{\ln l_0 - a_{0\nu}^n \ln a}{\ln l_0 - \ln a} + \frac{a_{0\nu}^n - 1}{\ln l_0 - \ln a} \ln r$$

$$+ \beta \sum_{k=1}^{n} (u_\nu^k - a_{0\nu}^n) R_0(r, \theta, \theta_k), \qquad (20.8)$$

$$v_{n\nu}(r,\theta) = C + \frac{a_{0\nu}^n - 1}{\ln l_0 - \ln a}\theta + \beta \sum_{k=1}^{n} (u_\nu^k - a_{0\nu}^n) I_0(r, \theta, \theta_k), \qquad (20.9)$$

where R_0, I_0 are kernels (20.4), C is an arbitrary constant,

$$a_{0\nu}^n = \frac{1}{n} \sum_{k=1}^{n} u_\nu^k.$$

In the process of solving system (20.7) we calculate only one control quantity

$$\varepsilon_{n\nu}^0 = \max_{1 \le p \le 21} |u_{n\nu}(A_p)|,$$

where A_p, $p = 1, 2, \ldots, 21$, are points lying on the line segment LF (see Fig. 32) with step $h = 1/20$. This quantity characterizes the deviation of function (20.8) from zero on the whole external boundary of domain (18.1) (with due account of its symmetry properties like those of the function u). The iterations in the solution of system (20.7) are terminated at the minimum $\nu \ge 2$, for which condition (15.13) is fulfilled. The approximate values ϑ_n and ρ_{1n}, M_n of quantities (16.8) and (16.9) can be found from the formula

$$\vartheta_n = (\ln l_0 - \ln a)/(a_{0\nu}^n - 1)$$

and from formulas (17.15), respectively. For this ϑ_n the function $\vartheta_n v_{n\nu}(r, \theta)$ has an increment equal to -2π upon the traverse of the contour Γ_1, i.e., of the internal boundary of domain (18.1).

The approximate conformal mapping of domain (18.1) onto a ring

$$\rho_{1n} < |\zeta| < 1 \tag{20.10}$$

is defined by an analytic function that has expression (17.17), with $U_n = \vartheta_n u_{n\nu}$, $V_n = \vartheta_n v_{n\nu}$, where $u_{n\nu}$, $v_{n\nu}$ are functions (20.8), (20.9). Here the images of 160 points (the points A_p, $p = 1, 2, \ldots, 21$, inclusive), lying on the external boundary Γ_0 of domain (18.1) with step $h = 1/20$, deviate from the circle $|\zeta| = 1$, i.e., from the external boundary of ring (20.10), by not more than quantity (17.19). The internal boundary of domain (18.1), i.e., the circle $|z| = a$, is mapped onto the internal boundary of ring (20.10), i.e., on the circle $|\zeta| = \rho_{1n}$.

Table 20.1 reflects the results of the construction on the BESM-6 computer of an approximate conformal mapping of domain (18.1) onto a ring for two values of a. For simplicity we have presented the difference between the conformal invariant M of domain (18.1) from Table 18.1 and its approximate value M_n obtained according to our algorithm.

The quantity $M - M_n$ lies in the limits of accuracy of the representation of M_n on the BESM-6, i.e., does not exceed $2^{-41} M_n$.

21. A Block-Bridge

In Sec. 18 and 20 we presented two variants of an approximate conformal mapping of domain (18.1), i.e., of a square with a circular hole, onto a ring by the block method. In both variants upper restrictions are imposed on the radius a of the hole in domain (18.1). When a circular lune block is used, the accuracy of the result decreases with a decrease in a, apparently because of the increased nonuniformity in the distribution of the points P'_k, $k = 1, 2, \ldots, n$, on the boundary of

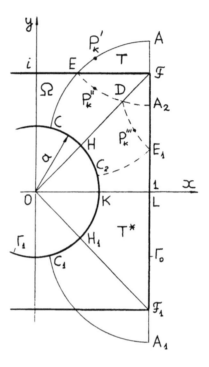

Figure 33

this block. However, the use of a block-ring is effective even for small values of a (see Sec. 22).

In this section we consider a third variant of an approximate conformal mapping of domain (18.1) onto a ring, a variant that is effective when

$$0.2 \leq a < 1,$$

which includes values of a close to 1 (see Sec. 22).

Fig. 33 presents the right half of domain (18.1) denoted by Ω. We assume the circular quadrangle HFF_1H_1, which constitutes a quarter of the domain Ω, to be the basic block T^* and place it in a right-angled circular quadrangle CAA_1C_1 symmetric about the real axis. We call the circular quadrangle CAA_1C_1 a block-bridge (an extended block) and denote it by T. The arc CA is defined by its distance τ from the point $z_0 = 2i$ on the imaginary axis, $a \leq \tau \leq 1$.

The center of the circle containing the arc CA lies on the straight line LF and has an ordinate

$$\varkappa_0 = (q - (q^2 - p(q^2 - 4\tau^2(1 - a^2))/4)^{1/2})/p,$$

where $p = 4 - \tau^2$, $q = a^2 + p$.

Let us consider an auxiliary function

$$w = w(z) = \exp\left\{\frac{i\pi}{\ln w(a)} \ln \omega(z)\right\}, \tag{21.1}$$

where

$$\omega(z) = \frac{z - 1 + (1 - a^2)^{1/2}}{1 + (1 - a^2)^{1/2} - z}.$$

Function (21.1) defines a conformal mapping of the block T onto a half-ring

$$l_1 < |w| < l_0, \qquad \mathrm{Im}\, w > 0, \tag{21.2}$$

where

$$l_0 = \exp\left\{\frac{\pi}{|\ln w(a)|}\left(\frac{\pi}{2} + \tan^{-1}\frac{\varkappa_0}{(1 - a^2)^{1/2}}\right)\right\},$$

$l_1 = 1/l_0$. Here the segment $A_1 A$ and the arc CC_1 (Fig. 33) pass into respective intervals $[l_1, l_0]$ and $[-l_0, -l_1]$ of the real axis in the complex plane w and the arcs CA and $C_1 A_1$ are mapped, in turn, onto semicircles $|w| = l_0$, $\mathrm{Im}\, w \geq 0$ and $|w| = l_1$, $\mathrm{Im}\, w \geq 0$.

We denote by $z(w)$ the elementary function that is the inverse of (21.1) and by $u(z)$ the harmonic extension (by the technique indicated in Sec. 20) of the solution of the corresponding boundary-value problem (16.6) on the domain Ω to the block-bridge T. Also suppose that $r = |w|$, $\theta = \arg w$, $\xi(w)$ is a function, harmonic on the half-ring (21.2), that assumes the following values on its boundary:

$$\xi = \begin{cases} 0, & \mathrm{Im}\, w = 0,\ l_1 \leq \mathrm{Re}\, w \leq l_0, \\ 1, & \mathrm{Im}\, w = 0,\ -l_0 \leq \mathrm{Re}\, w \leq -l_1, \\ u(z(w)), & (|w| - l_0)(|w| - l_1) = 0. \end{cases} \tag{21.3}$$

Obviously, relations of the form (18.6) are satisfied on the closed block $\overline{T}$ being considered.

By virtue of (21.1), (21.3) and the evenness of the function u relative to the real axis, we have

$$\xi(l_0 e^{i\eta}) = \xi(l_1 e^{i\eta}), \qquad 0 \leq \eta \leq \pi. \tag{21.4}$$

We set

$$\frac{R_0}{I_0}(r,\theta,\eta) = \sum_{p=0}^{1}\sum_{q=0}^{1}(-1)^{p+q}\frac{\text{Re}}{\text{Im}}\ \zeta_0\Big(\frac{r}{l_p}\psi_q(\theta,\eta),h\Big), \qquad (21.5)$$

where ζ_0 is function (19.9), h is quantity (19.6),

$$\psi_q(\theta,\eta) = \exp\{i(\theta - (-1)^q\eta)\}.$$

Using the odd extension of the function $\xi - \theta/\pi$ across the real axis (in the complex plane w) onto the ring $l_1 < |w| < l_0$ and taking (21.4) into account, we get by (19.4) the following representation of the function ξ on the half-ring (21.2):

$$\xi(w) = \theta/\pi + \int_0^\pi (\xi(l_0 e^{i\eta}) - \eta/\pi)R_0(r,\theta,\eta)\,d\eta. \qquad (21.6)$$

We choose a natural number n and specify points P_k, $k = 1,2,\ldots,n$, with polar coordinates $r = r_k = l_0$, $\theta = \theta_k = (k - 1/2/)\beta$, on the boundary of half-ring (21.2) with a constant angular step

$$\beta = \pi/n. \qquad (21.7)$$

We denote by P_k' the inverse image of the point P_k on the arc CA under mapping (21.1) (Fig. 33). With every point P_k' on the arc EA we associate a point P_k'' on the arc EA_2, symmetric to it about the straight line EF. If the point P_k' is on the arc CE, then let $P_k'' = P_k'$. With every point P_k'' lying on the union of the arcs CE and ED we associate a point $P_k''' \in C_2E_1 \cup E_1D$, symmetric to it about the straight line HF. If the point P_k'' belongs to the arc DA_2, then $P_k''' = P_k''$. Finally, we denote by P_k^0, $k = 1,2,\ldots,n$, the image of the point P_k''' under mapping (21.1). This image lies in the half-ring (21.2) or on the part of its boundary lying on the real axis.

Since the funcion u is even relative to the straight line HF, its extension across the straight line EF is odd and the functions u and ξ satisfy relations of the form (18.6), we have

$$\xi(P_k) = \begin{cases} -\xi(P_k^0), & P_k' \in EA, \\ \xi(P_k^0), & P_k' \in CA \setminus EA. \end{cases} \qquad (21.8)$$

Let $r = r_k^0$, $\theta = \theta_k^0$ be the polar coordinates of the point P_k^0, $k = 1, 2, \ldots, n$, in the complex plane w. Approximating the integral in (21.6) by the quadrature formula of rectangles (with nodes P_m, $m = 1, 2, \ldots, n$) and using relations (21.8), we arrive at a system of linear algebraic equations for the approximate values ξ^k of the function ξ at the points P_k (i.e., in the nodes), or, what is the same, for the approximate values $u^k = \xi^k$ of the function u at the points P_k'. This system has the form

$$u^k = \sigma_k\left(\theta_k^0/\pi + \beta \sum_{m=1}^{n}(u^m - \theta_m/\pi)R^{km}\right), \qquad 1 \le k \le n, \quad (21.9)$$

where β is angular step (21.7), $\theta_m = (m - 1/2)\beta$,

$$R^{km} = R_0(r_k^0, \theta_k^0, \theta_m),$$

$$\sigma_k = \begin{cases} -1, & P_k' \in EA, \\ 1, & P_k' \in CA \setminus EA. \end{cases}$$

Suppose that u_ν^k, $k = 1, 2, \ldots, n$, is the approximate solution of system (21.9) obtained after the νth iteration in the sense of Seidel ($u_0^k = \sigma_k$, $k = 1, 2, \ldots, n$). The approximate solution $u_{n\nu}(z)$ of the corresponding problem (16.6) on the circular quadrangle HFF_1H_1 (Fig. 33) and its conjugate harmonic function $v_{n\nu}(z)$ have the form

$$u_{n\nu}(z) = \frac{\arg w(z)}{\pi} + \beta \sum_{k=1}^{n}(u_\nu^k - \frac{\theta_k}{\pi})R_0(|w(z)|, \arg w(z), \theta_k), \quad (21.10)$$

$$v_{n\nu}(z) = -\frac{\ln |w(z)|}{\pi} + \beta \sum_{k=1}^{n}(u_\nu^k - \frac{\theta_k}{\pi})I_0(|w(z)|, \arg w(z), \theta_k), \quad (21.11)$$

where R_0, I_0 are kernels defined in (21.5), β is angular step (21.7), $\theta_k = (k - 1/2)\beta$.

Let B_q, $q = 1, 2, \ldots, 31$, be points arranged on the line segment HF (Fig. 33) with step $(\sqrt{2} - a)/30$ and numbered in order, the point B_1 coinciding with H.

Table 21.1

a	τ	n	ν	$M - M_n$	E_n^1
0.2	0.80	25	7	$1 \cdot 10^{-11}$	$1.3 \cdot 10^{-11}$
0.4	0.75	16	8	$1 \cdot 10^{-12}$	$1.2 \cdot 10^{-11}$
0.6	0.82	14	6	$-6 \cdot 10^{-12}$	$7.5 \cdot 10^{-12}$
0.8	0.90	11	6	$1 \cdot 10^{-12}$	$5.3 \cdot 10^{-12}$

When solving system (21.9) we compute only one control parameter $\varepsilon_{n\nu}^1$ from formula (17.13). Seidel's iterations are terminated at the minimum $\nu \geq 2$ for which condition (15.13) holds with the replacement of ε_{np}^0 by ε_{np}^1, $p = \nu - 1, \nu$.

Since $v_{n\nu} \equiv 0$ on KL, we can find the approximate values ϑ_n and ρ_{1n}, M_n of quantities (16.8) and (16.9) from the formula

$$\vartheta_n = \pi/4v_{n\nu}^0$$

and formulas (17.15), respectively.

The approximate conformal mapping of the circular quadrangle HFF_1H_1 (Fig. 33) onto a quarter of the ring (17.16) is defined by a function of the form (17.17), where U_n and V_n are defined by formulas (17.18) in terms of functions (21.10), (21.11). The arguments of the images of the points B_q, $q = 1, 2, \ldots, 31$, deviate from $\pi/2$ not more than quantity (17.20). The images of the line segment F_1F and of the arc H_1H (Fig. 33) lie on the circles $|\zeta| = 1$ and $|\zeta| = \rho_{1n}$, respectively.

Table 21.1 presents the results of computations on the BESM-6 computer for four values of a according to the algorithm given above. Here the quantity M for $a = 0.2$, $a = 0.4$, and $a = 0.8$ is taken from Table 18.1, and M_{70} from Table 18.3 is used for M when $a = 0.6$.

For $a = 0.6$ the quantity $|M - M_n|$ constitutes roughly four units in the least significant bit of the representation of M_n on the BESM-6 computer. In all other cases the quantity $|M - M_n|$ corresponds to a smaller number of unities in the least significant bit in the expression for M_n.

Recall that Tables 18.2, 18.3, 20.1, and 21.1 give the number n of the required real parameters and Table 18.1 indicates the number n_{opt} of the desired complex parameters.

22. Limit Cases

In this section we illustrate the great potential possibilities of the block method by using it in limited cases, namely, when a is small (domain (18.1) has a small hole) or when a is close to unity (domain (18.1) breaks up into four parts with narrow gaps between them). The variant of the method given in Sec. 20 is used for mapping for a small a and a block-bridge is used for a close to 1.

Let $w = f(z)$ be a function mapping conformally the square $\{z : |x| < 1, |y| < 1\}$ onto a ring $|w| < 1$, with $f(0) = 0$. When domain (18.1) is conformally mapped onto a ring formed by concentric circles of radii ρ_0 and ρ_1, $\rho_1 < \rho_0 = 1$, it is natural to expect that the ratio $\varkappa = \rho_1/a$ tends to the quantity $\lambda_0 = |f'(0)|$ as $a \to +0$. This conjecture receives the following confirmation in practice.

The approximate value λ_0^* of the quantity λ_0,

$$\lambda_0^* = 0.927\ 0373\ 3865$$

was found by means of an approximate conformal mapping of a square onto the strip $|\operatorname{Re} \zeta| < \pi/4$ constructed according to the algorithm given in 15.2 and the mapping $t = \tan \zeta$ of the indicated strip on the unit disk $|t| < 1$.

Let $\varkappa_n = \rho_{1n}/a$, where ρ_{1n} is an approximate value of the internal radius of the ring calculated from the first formula of (17.15). For $n = 20$ the algorithm presented in Sec. 20 was used to obtain the results presented in Table 22.1.

The value $a = 10^{-18}$ is close to the minimum number different from zero on the BESM-6, and this apparently caused an (additional)

Table 22.1

a	l_0	ν	$\varkappa_n$	E_n^0
0.4	1.60	11	0.927 05	$2.5 \cdot 10^{-9}$
0.2	1.70	12	0.927 0374	$2.2 \cdot 10^{-11}$
10^{-1}	1.75	15	0.927 0373 39	$1.7 \cdot 10^{-11}$
10^{-3}	1.75	16	0.927 0373 3865	$1.8 \cdot 10^{-11}$
10^{-12}	1.75	13	0.927 0373 3864	$2.3 \cdot 10^{-11}$
10^{-18}	1.75	13	0.927 0373 3868	$2.5 \cdot 10^{-11}$

error of 3 units in the eleventh digit of $\varkappa_n$.

Of interest in the second limit case, as $a \to 1$, is the asymptotic behavior

$$1 - \rho_1 \sim \delta(a), \qquad 1 - a \to +0, \tag{22.1}$$

where $1 - \rho_1$ is the width of the ring onto which domain (18.1) is mapped,

$$\delta(a) = ((1 - a)/8)^{1/2}.$$

The derivation of the asymptotic expression (22.1) is omitted. We set

$$\omega_n = (1 - \rho_{1n})/\delta(a),$$

where ρ_{1n} is the approximate value of the internal radius of the ring (the external radius is equal to unity) onto which domain (18.1) is mapped, computed by the method presented in Sec. 21. The results for $\tau = a$ computed on the BESM-6 (see Sec. 21) are given in Table 22.2.

Table 22.2

$1-a$	n	ν	ω_n	E_n^1
10^{-2}	10	5	1.10	$8.1 \cdot 10^{-12}$
10^{-5}	8	5	1.003	$7.4 \cdot 10^{-12}$
10^{-10}	8	4	1.000 009	$3.4 \cdot 10^{-12}$
10^{-12}	6	3	1.000 000 07	$9.5 \cdot 10^{-12}$

Table 22.1 and 22.2 testify to the high accuracy and effectiveness of the block method when the domain has elements of fine structure (a small hole, narrow gaps). Here n (the number of the required real parameters) and ν (the number of iterations performed) assume small values.

23. Mapping a Disk with an Elliptic Hole or with a Retrosection onto a Ring

Let $z = x + iy$, $0 \le b < a$, $c = (a^2 - b^2)^{1/2}$, $1 + c^2 < 1/a^2$,

$$\Omega = \begin{cases} \{z : |z| < 1\} \setminus \{z : \frac{x^2}{a^2} + \frac{y^2}{b^2} \le 1\}, & b > 0, \\ \{z : |z| < 1\} \setminus \{z : |x| \le a, y = 0\}, & b = 0, \end{cases} \tag{23.1}$$

$$\Gamma_0 = \{z : |z| = 1\}, \qquad \Gamma_1 = \overline{\Omega} \setminus (\Omega \cup \Gamma_0).$$

We place the given domain Ω in the domain

$$D = \left\{z : \frac{x^2}{A^2} + \frac{y^2}{B^2} < 1\right\} \cap \{z : |z| \ge 1\} \cup \Omega,$$

where A is chosen to satisfy the inequalities

$$(1 + c^2)^{1/2} < A < 1/a,$$

and

$$B = (A^2 - c^2)^{1/2} > 1.$$

Obviously, the external boundary of the doubly-connected domain D is an ellipse, which has foci in common with the ellipse Γ_1 when $b > 0$ and its foci are the endpoints of the given retrosection Γ_1 when $b = 0$.

On $\overline{D} \setminus \overline{\Omega}$ we complete the definition of the function $u(z)$, which is the solution of the boundary-value problem (16.6) on domain (23.1), as follows:

$$u(z) = -u(z/|z|^2), \qquad z \in \overline{D} \setminus \overline{\Omega}. \tag{23.2}$$

Obviously, the function $u(z)$ defined in this way is harmonic on the domain D.

Let us consider an auxiliary function

$$w = w(z) = z/c + ((z/c)^2 - 1)^{1/2}, \tag{23.3}$$

mapping conformally the domain D onto a ring

$$K = \{w : l_1 < |w| < l_0\}, \tag{23.4}$$

where $l_0 = A/c + ((A/c)^2 - 1)^{1/2}$, $l_1 = (a + b)/c$.

Suppose that $z = z(w)$ is the inverse of (23.3), $\xi(w)$ is a function harmonic on the ring K and assuming the following values on its boundary:

$$\xi = \begin{cases} 1, & |w| = l_1, \\ u(z(w)), & |w| = l_0. \end{cases}$$

Then, evidently,

$$\xi(w) = u(z(w)), \qquad w \in \overline{K}, \tag{23.5}$$

$$u(z) = \xi(w(z)), \qquad z \in \overline{D}. \tag{23.6}$$

We set $r = |w|$, $\theta = \arg w$. We specify the points P_k, $k = 1, 2, \ldots, n$, with polar coordinates $r = r_k = l_0$, $\theta = \theta_k = (k - 1/2)\beta$, where $\beta = \pi/2n$ on the part of the external boundary of the ring

K lying in the first quadrant. Moreover, let P'_k, $k = 1, 2, \ldots, n$, be points which have polar coordinates

$$
\begin{aligned}
r &= r'_k = |w(z(w_k)/|z(w_k)|^2)|, \\
\theta &= \theta'_k = \arg w(z(w_k)/|z(w_k)|^2),
\end{aligned}
\tag{23.7}
$$

where

$$
w_k = l_0 \exp\{i\theta_k\}.
$$

The points P'_k, $k = 1, 2, \ldots, n$, lie strictly inside the ring (23.4) at a distance greater than a certain positive quantity, independent of n, from its external boundary. This condition corresponds to the requirement formulated in Remark 6.4.

By virtue of (23.2), (23.5)–(23.7) we have

$$
\xi(P_k) = -\xi(P'_k), \qquad k = 1, 2, \ldots, n.
\tag{23.8}
$$

Relations (23.8) play the same part as (18.8), (20.3), (21.8) in setting up the appropriate system of algebraic equations. The algorithm of approximation of the function ξ and its conjugate function on the domain that is the image of Ω under mapping (23.3) is similar to the corresponding algorithm from Sec. 20 for obtaining an approximate solution of problem (16.6) on domain (18.1) and its conjugate function. The difference is mainly that in the case considered in Sec. 20 the required function u has four axes of symmetry whereas in this case the function ξ has only two symmetric axes, the real and the imaginary axis. A change of the independent variable of form (23.6) in the approximate expressions found for the function ξ and its conjugate yields an approximate solution of the original boundary-value problem (16.6) on domain (23.1) and its conjugate function (a similar change of the independent variable is carried out in Sec. 18 and Sec. 21). In the computation process we take as the control parameter $\varepsilon^0_{n\nu}$ the maximum deviation of the approximate solution of problem (16.6) from zero at 12 points arranged on the arc $|z| = 1$, $\pi/48 \le \arg z \le \pi/2 - \pi/48$, with the step $\beta' = \pi/24$.

We present the results of the computation carried out on the BESM-6 computer for the most interesting case $b = 0$, i.e., for an approximate conformal mapping of a disk with an interior cut onto a

Table 23.1

a	ν	ρ_{1n}	E_n^0
0.7	8	0.3619	$9.4 \cdot 10^{-5}$
0.6	10	0.3051 82	$7.0 \cdot 10^{-7}$
0.5	13	0.2520 127	$9.9 \cdot 10^{-9}$
0.4	16	0.2006 4780 5	$2.5 \cdot 10^{-10}$
0.3	17	0.1501 5245 46	$1.1 \cdot 10^{-11}$
0.2	19	0.1000 2001 50	$5.1 \cdot 10^{-12}$
0.1	18	0.0500 0062 50	$7.8 \cdot 10^{-12}$

ring. In this case domain (23.1) has reentral angles equal to 2π and, consequently (see [50]), even the first derivatives of the function u are unbounded in the neighborhoods of the vertices of these angles, i.e., the ends of the cut. However, this does not disturb the exponential behavior of the convergence of the block method. The results corresponding to the case $b = 0$, $A = 1.3$, $n = 25$ are presented in Table 23.1, where a is half the length of the interior cut, ν is the number of iterations needed to solve the corresponding system of linear algebraic equations, ρ_{1n} is the approximate value of the internal radius of the ring onto which domain (23.1) is mapped, E_n^0 is the maximum deviation of the approximate images of the indicated 12 points, lying on the external boundary of domain (23.1), from the circle $|\zeta| = 1$.

The asymptotic behavior

$$\rho_{1n} \sim a/2, \qquad a \to +0,$$

which holds for ρ_1, finds clear expression in Table 23.1.

24. Mapping a Disk with a Regular Polygonal Hole

Here is an algorithm of an approximate conformal mapping of a doubly-connected domain onto a ring with the use of two pairs of blocks.

Suppose that Ω is a doubly-connected domain in the complex plane z, $z = x + iy$, its external boundary Γ_0 coincides with the circle $|z| = 2$, and its internal boundary Γ_1 is a regular t-sided closed polygonal line whose center is at the point $z = 0$ and one of the sides lies in the right half-plane and is orthogonal to the real axis. Fig. 34 shows the right side of the domain Ω in the case when $t = 3$. We understand P_1, P_3 to be the endpoints of the indicated side of the polygonal line. The arguments that follow for $t \geq 3$ are the same.

The domain Ω and the solution u of the boundary-value problem (16.6) on this domain have t axes of symmetry arranged with an angular step π/t. It is therefore sufficient to find the solution u and its conjugate harmonic function v on the circular quadrilateral AP_2FP_1 (Fig. 34) constituting a fraction $1/2t$ of the domain Ω. In what follows we denote by Ω_0 the open quadrilateral AP_2FP_1 and by $\overline{\Omega}_0$ its closure.

Let r_1, θ_1 and r_2, θ_2 be polar coordinate systems with poles P_1 and P_2, the angle θ_1 being reckoned from the ray P_1P_3, and θ_2 from the positive direction of the real axis. We are given extended blocks (a sector and a half-disk)

$$T_\mu = \{(r_\mu, \theta_\mu) : 0 < r_\mu < r_{\mu 0}, \, 0 < \theta_\mu < \alpha_\mu \pi\}$$

and basic blocks

$$T_\mu^* = \{(r_\mu, \theta_\mu) : 0 < r_\mu < r_\mu^*, \, 0 < \theta_\mu < \alpha_\mu \pi\},$$

which make a part of them, $r_\mu^* < r_{\mu 0}$, $\mu = 1, 2$, $\alpha_1 = 1/2 + 1/t$, $\alpha_2 = 1$. We assume that the relative position of the blocks are as in Fig. 34, where the point L_2 may coincide with the point A. In particular, $\Omega_0 \subset \overline{T}_1^* \cup \overline{T}_2^*$. Moreover, for any point $P \in \overline{T}_1 \setminus \overline{\Omega}$ the conjugate point relative to the circle Γ_0 lies in Ω.

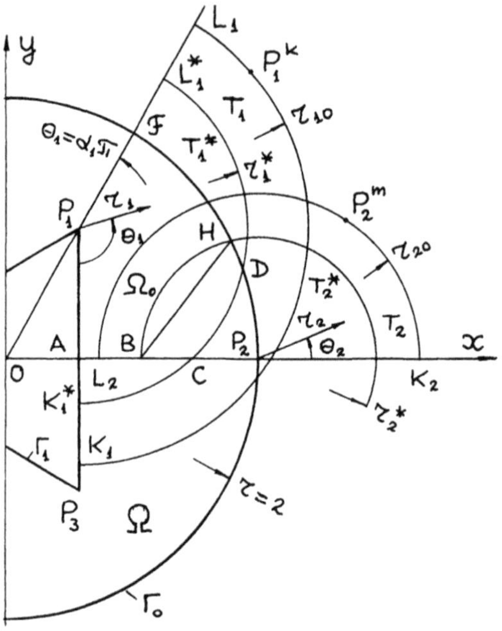

Figure 34

These requirements imposed on the blocks are met if the polygonal hole is not too small and not too large. Specific cases for the
indicated arrangement of the blocks are given below for $t = 3, 4, 6$
whereas a detailed discussion of the general case is outside of the
scope of this section. A suitable covering of the quadrilateral Ω_0
can be constructed for a hole of any size by increasing the number
of blocks or using blocks of other types (see, in particular, Sec. 18,
Sec. 20 and [67]).

Suppose, furthermore, that

$$\frac{R_1}{I_1}(r, \theta, \eta) = \frac{1}{2\alpha_1} \sum_{q=0}^{1} \sum_{p=0}^{1} (-1)^p \frac{R}{I}\left(\left(\frac{r}{r_{10}}\right)^{1/2\alpha_1}, \frac{\theta}{2\alpha_1}, \varphi_{pq}(\eta)\right),$$

$$\frac{R_2}{I_2}(r, \theta, \eta) = \sum_{p=0}^{1} \frac{R}{I}\left(\frac{r}{r_{20}}, \theta, (-1)^p \eta\right),$$

where $\varphi_{pq}(\eta) = (-1)^p(\pi/2 - (-1)^q(\pi/2 - \eta))/2\alpha_1$, $\alpha_1 = 1/2 + 1/t$, and R, I are the kernel of Poisson's integral (3.14) and its conjugate kernel (8.13). Since the solution u of problem (16.6) vanishes on Γ_0, it can be easily continued analytically (harmonically) from Ω across Γ_0, in particular, onto $(\overline{T}_1 \cup \overline{T}_2) \setminus \overline{\Omega}$. We retain the same notation u for the extended function. The function u is even with respect to the straight line OL_1 and relative to the real axis and is obviously continuous on the closed basic blocks $\overline{T}_1^*$, $\overline{T}_2^*$. Consequently, on the basis of Theorem 3.1, this function can be represented as

$$u(r_1, \theta_1) = 1 + \int_0^{\alpha_1 \pi} (u(r_{10}, \eta) - 1)R_1(r_1, \theta_1, \eta)\, d\eta \quad \text{on} \quad \overline{T}_1^*, \quad (24.1)$$

$$u(r_2, \theta_2) = \int_0^{\pi} u(r_{20}, \eta)R_2(r_2, \theta_2, \eta)\, d\eta \quad \text{on} \quad \overline{T}_2^*. \quad (24.2)$$

We choose a natural number n and specify, on the arcs $K_\mu L_\mu$ (Fig. 34), $\mu = 1, 2$, points P_μ^k, $k = 1, 2, \ldots, n$, with polar coordinates $r_\mu = r_\mu^k = r_{\mu 0}$, $\theta_\mu = \theta_\mu^k = (k - 1/2)\beta_\mu$, where $\beta_\mu = \alpha_\mu \pi/n$ is the angular step ($\alpha_1 = 1/2 + 1/t, \alpha_2 = 1$). We use the following rule to put a point $P_\mu^{k0} \in \overline{\Omega}_0$ into correspondence with every point P_μ^k. Let $z_\mu^k = x_\mu^k + iy_\mu^k$ be a complex coordinate of the point P_μ^k. Then

$$z_\mu^{k0} = \begin{cases} x_\mu^k + i|y_\mu^k|, & P_\mu^k \in \overline{\Omega}, \\ 4(x_\mu^k + i|y_\mu^k|)/|z_\mu^k|^2, & P_\mu^k \notin \overline{\Omega}, \end{cases}$$

is a complex coordinate of the point P_μ^{k0}. Obviously,

$$u(P_\mu^k) = \begin{cases} u(P_\mu^{k0}), & P_\mu^k \in \overline{\Omega}, \\ -u(P_\mu^{k0}), & P_\mu^k \notin \overline{\Omega}. \end{cases} \quad (24.3)$$

Approximating the integrals in (24.1) and (24.2) by the quadrature formulas of rectangles (with nodes P_1^m and P_2^m, $m = 1, 2, \ldots, n$) and taking into account (24.3), we arrive, as in Sec. 15, 17, 18, 21–23, at the following system of linear algebraic equations for the approximate values u_μ^k of the function u at the points P_μ^k:

$$u_\mu^k = \sigma_\mu^k \begin{cases} \left(1 + \beta_1 \sum_{m=1}^{n}(u_1^m - 1)R_{1\mu}^{km}\right), & P_\mu^{k0} \in \overline{T}_1^*, \\[4mm] \beta_2 \sum_{m=1}^{n} u_2^m R_{2\mu}^{km}, & P_\mu^{k0} \in \overline{T}_2^* \setminus \overline{T}_1^*, \end{cases} \tag{24.4}$$

$$1 \le \mu \le 2, \qquad 1 \le k \le n,$$

where $\beta_\tau = \alpha_\tau \pi / n$, $\tau = 1, 2$, $\alpha_1 = 1/2 + 1/t$, $\alpha_2 = 1$,

$$\sigma_\mu^k = \begin{cases} 1, & P_\mu^k \in \overline{\Omega}, \\ -1, & P_\mu^k \notin \overline{\Omega}, \end{cases}$$

$R_{\tau\mu}^{km} = R_\tau(r_{\tau\mu}^k, \theta_{\tau\mu}^k, \theta_\tau^m)$, $\theta_\tau^m = (m - 1/2)\beta_\tau$, and $r_{\tau\mu}^k$, $\theta_{\tau\mu}^k$ are the polar coordinates of the point P_μ^{k0} in the coordinate system r_τ, θ_τ. Here the normalizing denominator, appearing in (4.7), which is close to unity, is omitted for the sake of simplicity in the calculations of the coefficients $R_{\tau\mu}^{km}$. This is permissible when the number of blocks is small (two pairs in this case).

Let $u_{\nu\mu}^k$, $1 \le \mu \le 2$, $1 \le k \le n$, be an approximate solution of system (24.4), obtained after the νth Seidel iteration for the initial approximation $u_{0\mu}^k = \sigma_\mu^k/2$. On the extended blocks T_1 and T_2 the following pairs of conjugated harmonic functions are defined:

$$\begin{matrix} u_{1n}^\nu \\ v_{1n}^\nu \end{matrix} (r_1, \theta_1) = \begin{matrix} 1 \\ 0 \end{matrix} + \beta_1 \sum_{k=1}^{n} (u_{\nu 1}^k - 1) \begin{matrix} R_1 \\ I_1 \end{matrix} (r_1, \theta_1, \theta_1^k),$$

$$\begin{matrix} u_{2n}^\nu \\ v_{2n}^\nu \end{matrix} (r_2, \theta_2) = \beta_2 \sum_{k=1}^{n} u_{\nu 2}^k \begin{matrix} R_2 \\ I_2 \end{matrix} (r_2, \theta_2, \theta_2^k),$$

where $\beta_\mu = \alpha_\mu \pi / n$, $\theta_\mu^k = (k - 1/2)\beta_\mu$ $(\alpha_1 = 1/2 + 1/t, \alpha_2 = 1)$. These functions are continuous on the closed basic blocks $\overline{T}_1^*$ and $\overline{T}_2^*$, respectively.

The function $u_{\mu n}^\nu$, $\mu = 1, 2$, is the approximate solution of he boundary-value problem (16.6) on $\overline{T}_\mu^* \cap \overline{\Omega}_0$ (on the intersection of the closed basic block $\overline{T}_\mu^*$ and the closed circular quadrilateral AP_2FP_1).

The solution u of the problem (16.6) on the domain being considered and its conjugate function v (a single-valued branch) satisfy the following conditions (see Fig. 34):

$$u = 1 \quad \text{on} \quad AP_1, \qquad \partial u / \partial \theta_1 = 0 \quad \text{on} \quad P_1 F \setminus P_1, \tag{24.5}$$

$$u = 0 \quad \text{on} \quad P_2 H F, \tag{24.6}$$

$$v = c_1 \quad \text{on} \quad A P_2, \tag{24.7}$$

$$v = c_2 \quad \text{on} \quad P_1 F, \tag{24.8}$$

where c_1, c_2 are constants.

For any ν the function u_{1n}^ν also satisfies conditions (24.5) and its conjugate v_{1n}^ν satisfies condition (24.8) for $c_2 = 0$. The function v_{2n}^ν satisfies condition (24.7) on the line segment $BP_2 \subset AP_2$ for $c_1 = 0$. Condition (24.6) for u_{1n}^ν on the arc DHF and for u_{2n}^ν on the arc $P_2 DH$ is only approximately satisfied. Condition (24.7) for v_{1n}^ν on the line segment $AC \subset AP_2$ is also only approximately satisfied.

To control Seidel's iteration process in solving system (24.4) and to characterize the accuracy of the approximate solutions $u_{\mu n}^\nu$, $\mu = 1, 2$, we evaluate, after each iteration, the control quantity

$$\varepsilon_{n\nu}^0 = \max_{1 \le p \le 15} |u_n^\nu(A_p)|, \tag{24.9}$$

where

$$u_n^\nu(A_p) = \begin{cases} u_{1n}^\nu(A_p), & A_p \in \overline{T}_1^*, \\ u_{2n}^\nu(A_p), & A_p \in \overline{T}_2^* \setminus \overline{T}_1^*, \end{cases}$$

A_p is a point on the arc $P_2 H F$ which has a complex coordinate

$$z_p = 2 \exp\{i\pi(2p - 1)/30t\}.$$

Quantity (24.9) characterizes the deviation of the approximate solutions u_{1n}^ν and u_{2n}^ν from zero on the arcs DHF and $P_2 D$, on which the solution u of problem (16.6) vanishes. The iterations are terminated at the minimum $\nu \ge 2$, for which inequality (15.13) is satisfied.

After finishing iterations, we calculate three more control parameters ε_n^1, ε_n^2, ε_n^3 as follows. Let B_q (C_q, respectively), $q = 1, 2, \ldots, 12$, be points lying on the line segment AC (on the line segment BH), with a constant step, the point B_1 (C_1, respectively) coinciding with A (B), and the point B_{12} (C_{12}) with C (H). Then

$$\varepsilon_n^1 = \max_{1 \le q \le 12} |v_{1n}^\nu(B_q) - w_{1n}|,$$

$$\varepsilon_n^2 = \max_{1 \le q \le 12} |v_{1n}^\nu(C_q) - v_{2n}^\nu(C_q) - w_n|,$$

$$\varepsilon_n^3 = |w_{1n} - w_n|,$$

where

$$w_{1n} = \frac{1}{12} \sum_{q=1}^{12} v_{1n}^\nu(B_q),$$

$$w_n = \frac{1}{12} \sum_{q=1}^{12} (v_{1n}^\nu(C_q) - v_{2n}^\nu(C_q)).$$

The quantity ε_n^1 is equal to the maximum deviation of the function v_{1n}^ν (conjugate to the approximate solution u_{1n}^ν) at 12 points of the line segment AC from its mean value w_{1n} over these points. The quantity ε_n^2 is the maximum deviation of the difference $v_{1n}^\nu - v_{2n}^\nu$ from its mean value at 12 points of the line segment BH. Finally, the quantity ε_n^3 can be regarded as the difference of the mean values of the difference $v_{1n}^\nu - v_{2n}^\nu$ on BH and AC over 12 points since $v_{2n}^\nu \equiv 0$ on the real axis.

The approximate values ϑ_n and ρ_{1n}, M_n of quantities (16.8) and (16.9) can be found from the formula

$$\vartheta_n = -\pi/tw_{1n}$$

and formulas (17.15), respectively, since $v_{1n}^\nu \equiv 0$ on $P_1 F$ and, consequently, $w_{1n} - 0 \approx \delta v/2t$.

The approximate conformal mapping of the open quadrangle AP_2FP_1 (the fraction $1/2t$ of the given domain Ω) onto the fraction $1/2t$ of the ring

$$\rho_{1n} < |\zeta| < 1, \qquad 0 < \arg \zeta < \pi/t$$

is specified by parts, namely,

$$\zeta = \zeta_{1n}(z) = \exp\{\vartheta_n(u_{1n}^\nu + iv_{1n}^\nu) + i\pi/t\} \quad \text{on} \quad \overline{T}_1^* \cap \Omega_0,$$

$$\zeta = \zeta_{2n}(z) = \exp\{\vartheta_n(u_{2n}^\nu + iv_{2n}^\nu)\} \quad \text{on} \quad (\overline{T}_2^* \setminus \overline{T}_1^*) \cap \Omega_0,$$

and both functions ζ_{1n} and ζ_{2n} are defined on $\overline{T}_1^* \cap \overline{T}_2^* \cap \Omega_0$, where they assume close values. The images of the points A_p, $p = 1, 2, \ldots, 15,$

on the arc P_2HF deviate, under the mapping we have constructed, from the circle $|\zeta| = 1$ by not more than the quantity

$$E_n^0 = \exp\{|\vartheta_n|\varepsilon_{n\nu}^0\} - 1,$$

and the arguments of the images of the points B_q, $q = 1, 2, \ldots, 12$, on AC deviate from zero by not more than

$$E_n^1 = |\vartheta_n|\varepsilon_n^1.$$

The image of the line segment AP_1 lies entirely on the circle $|\zeta| = \rho_{1n}$. The small quantity $E_n^2 = |\vartheta_n| \max\{\varepsilon_n^2, \varepsilon_n^3\}$ characterizes the closeness of the functions $\vartheta_n v_{1n}^\nu + \pi/t$ and $\vartheta_n v_{2n}^\nu$ in their common domain of definition $\overline{T}_1^* \cap \overline{T}_2^* \cap \overline{\Omega}_0$.

Our approximate conformal mapping of the subdomain Ω_0 (the circular quadrangle AP_2FP_1) can be extended to $\Omega \setminus \overline{\Omega}_0$ with the aid of the obvious properties of symmetry and the periodicity of function (16.5) in the case being considered.

Table 24.1 gives single-precision data calculated according to our algorithm by the BESM-6 computer (to 12-13 decimal places) for $t = 3, 4, 6$. Let x_A be the abscissa of the point A (see Fig. 34). We have chosen the following parameter values:

$$t = 3, \quad x_A = 1/\sqrt{3}, \quad r_{10} = 1.75, \quad r_1^* = 1.4, \quad r_{20} = 0.9, \quad r_2^* = 0.63,$$
$$t = 4, \quad x_A = 1, \quad r_{10} = 1.24, \quad r_1^* = 1.033, \quad r_{20} = 1, \quad r_2^* = 0.75,$$
$$t = 6, \quad x_A = 1.2, \quad r_{10} = 1.21, \quad r_1^* = 0.85, \quad r_{20} = 0.8, \quad r_2^* = 0.5.$$

In the table t is the number of sides of an equilateral polygon forming the hole in the disk $|z| < 2$; n is the value of the discreteness parameter, ν is the number of iterations carried out to solve system (24.4), M_n is the approximate value obtained for the conformal invariant of the domain being mapped, E_n^p, $p = 0, 1, 2$, are the indicators of the accuracy of the result introduced earlier.

The small values of ν testify to the rapid convergence of iterations when system (24.4) is being solved. The observed increase in the number of repeating digits in the expression for M_n substantiates the exponential convergence of the block method with respect to the parameter n. For a check, we quote below the value of the conformal invariant M of the domain Ω for $t = 4$, obtained in [34] by another

Table 24.1

t	n	ν	M_n	E_n^0	E_n^1	E_n^2
3	10	3	2.367	$7 \cdot 10^{-3}$	$3 \cdot 10^{-3}$	$4 \cdot 10^{-3}$
	20	7	2.366 53	$2 \cdot 10^{-5}$	$2 \cdot 10^{-5}$	$2 \cdot 10^{-5}$
	40	10	2.366 5296 8000	$2 \cdot 10^{-10}$	$1 \cdot 10^{-9}$	$4 \cdot 10^{-10}$
	50	10	2.366 5296 8009	$2 \cdot 10^{-11}$	$3 \cdot 10^{-11}$	$5 \cdot 10^{-11}$
4	10	4	1.691 8	$3 \cdot 10^{-3}$	$3 \cdot 10^{-3}$	$3 \cdot 10^{-3}$
	20	5	1.691 568	$3 \cdot 10^{-5}$	$3 \cdot 10^{-5}$	$2 \cdot 10^{-5}$
	40	11	1.691 5649 0261	$2 \cdot 10^{-10}$	$9 \cdot 10^{-10}$	$2 \cdot 10^{-9}$
	50	11	1.691 5649 0265	$6 \cdot 10^{-11}$	$4 \cdot 10^{-11}$	$4 \cdot 10^{-11}$
6	5	3	1.566	$3 \cdot 10^{-3}$	$2 \cdot 10^{-3}$	$2 \cdot 10^{-3}$
	15	7	1.568 0629	$1 \cdot 10^{-7}$	$5 \cdot 10^{-7}$	$4 \cdot 10^{-7}$
	30	10	1.568 0631 6002	$5 \cdot 10^{-11}$	$3 \cdot 10^{-10}$	$2 \cdot 10^{-10}$
	40	8	1.568 0631 6016	$7 \cdot 10^{-11}$	$1 \cdot 10^{-11}$	$6 \cdot 10^{-12}$

method on the CDC 7600 computer (which has more decimal places than he BESM-6 computer):

$$M = 1.691\ 5649\ 0259.$$

For $t = 4$ the value M_{40} is closer to M than the value M_{50}. This can be explained by the fact that for $n = 50$, and even for $n = 40$, the effect of the errors inherent in the block method is weaker than the effect of the BESM-6 rounding-off errors, whose influence increases with an increase in n (see Remark 7.7).

For practical purposes, it seems to suffice to set $n = 20$. It takes a few seconds to compute one variant on the BESM-6 computer.

25.   Mapping the Exterior of a Parabola with a Hole onto a Ring

This section shows the possibility of using the block method for an approximate conformal mapping of an infinite doubly-connected domain onto a ring.

We denote by Ω an infinte doubly-connected domain in the complex plane z, $z = x + iy$, bounded by a parabola $\Gamma_0 = \{z : y^2 = 4(1-x)\}$ and a circle $\Gamma_1 = \{z : |z-4| = a\}$, $0 < a \leq 2$. We construct the approximate mapping of the domain Ω onto a ring according to the scheme given in Sec. 16. We understand u to be a bounded solution of the boundary-value problem (16.6) on the infinite domain Ω being considered.

Let $D = \{z : y^2 > 4(1 - x)\}$ be the exterior of the parabola Γ_0, $z \in D$, $|\arg z| < \pi, \arg \sqrt{z} = (\arg z)/2$. We construct two extended blocks

$$T_1 = \{z : |z - b - 1| < b\} \setminus \{z : |z - 4| \leq a\},$$

$$T_2 = D \setminus \{z : |\sqrt{z} - \varkappa_0| \leq \omega_0\},$$

where $\varkappa_0 = (\sqrt{4 + a} + \sqrt{4 - a})/2$,

$$\omega_0 = ((\sqrt{4 + a} - \sqrt{4 - a})/2)^\alpha,$$

$1/2 \leq \alpha < 1, b > 3$.

We specify fractional-linear functions $w_1(z)$ and $w_2(z)$ such that $w_1(1) = w_2(1) = -1$, $w_1(2b + 1) = w_2(\infty) = 1$, $-w_1(4 - a) = w_1(4 + a) < 1$, $-w_2(\varkappa_0 - w_0) = w_2(\varkappa_0 + w_0) < 1$, and set

$$f_1(z) = w_1(z), \qquad f_2(z) = w_2(\sqrt{z}).$$

The function $w = f_\mu(z)$, $\mu = 1, 2$, conformally maps the extended block T_μ onto the ring $r_{\mu 0} < |w| < 1$, with $r_{10} = w_1(4 + a)$, $r_{20} = w_2(\varkappa_0 + w_0)$. We denote by $f_\mu^{-1}(w)$ the inverse of $f_\mu(z)$ and by γ a circle which is the image of some circle $|w| = r_{30} = r_{10}^\beta$, $0 < \beta < 1$, under the mapping $z = f_1^{-1}(w)$. We assume that the parameters b, α, and β are such that

$$\Omega = T_1 \cup T_2, \qquad \gamma \subset T_1 \cap T_2, \qquad \overline{T}_2 \cap \Gamma_1 = \varnothing. \qquad (25.1)$$

Below we indicate the specific values of the parameters b, α, and β for the values of a for which we have carried out the construction of the approximate conformal mapping of the domain Ω onto the ring.

We introduce basic blocks T_μ^*, $\mu = 1, 2$. The basic block T_μ^* is a part of the extended block T_μ bounded by the curve $\Gamma_{2-\mu}$ and the circle γ. Obviously, $T_1^* \cap T_2^* = \varnothing$, $\overline{T}_1^* \cap \overline{T}_2^* = \gamma$,

$$\overline{T}_1^* \cup \overline{T}_2^* = \overline{\Omega}, \qquad \overline{T}_\mu^* \cap \Omega \subset T_\mu, \qquad \mu = 1, 2. \qquad (25.2)$$

We set

$$\begin{matrix} R \\ I \end{matrix} (r, \theta, l, h, \eta) = \begin{matrix} \mathrm{Re} \\ \mathrm{Im} \end{matrix} \zeta_0(t, h), \qquad (25.3)$$

where $t = t(r, \theta, l, \eta)$ is function (19.8), $\zeta_0(t, h)$ is Weierstrass' function (19.9).

By virtue of Villat's formula (see [1] and also Sec. 19) and with due regard for the evenness of the, being considered, solution u of the boundary-value probem (16.6) relative to the real axis, we can represent it on the extended block T_μ, $\mu = 1, 2$, in the form

$$u(z) = a_{\mu 0} + (a_{\mu 1} - a_{\mu 0}) \frac{\ln |f_\mu(z)|}{\ln r_{\mu 0}} + \int\limits_0^\pi (\psi_\mu(\eta) - a_\mu) R_\mu(z, \eta) \, d\eta. \quad (25.4)$$

Here $a_{20} = 0$, $a_{11} = 1$,

$$R_\mu(z,\eta) = (-1)^{\mu-1} \sum_{q=0}^{1} R(|f_\mu(z)|, \arg f_\mu(z), g_\mu, r_{\mu 0}, (-1)^q \eta), \quad (25.5)$$

$$\psi_\mu(\eta) = u(f_\mu^{-1}(g_\mu e^{i\eta})), \quad (25.6)$$

$$a_\mu = a_{\mu,\mu-1} = \frac{1}{\pi} \int_0^\pi \psi_\mu(\eta) \, d\eta, \quad (25.7)$$

$$g_1 = 1, \qquad g_2 = r_{20}.$$

Thus we can express the solution u of the boundary-value problem (16.6) in form (25.4) throughout the extended block T_μ, $\mu = 1, 2$, in terms of its values on the boundary of the block. Here the part of the boundary of the extended block T_μ, on which the values of u are unknown, is inside the basic block $T_{3-\mu}^*$ (and this corresponds to Remark 6.4).

We choose a natural number n and specify points P_μ^k, $\mu = 1, 2$, $k = 1, 2, \ldots, n$, with a complex coordinate

$$z_\mu^k = f_\mu^{-1}(g_\mu \exp\{i\pi(k - 1/2)/n\}). \quad (25.8)$$

By virtue of (25.1), (25.2) we have

$$P_1^k \in T_2^* \subset T_2, \qquad P_2^k \in T_1^* \subset T_1, \qquad k = 1, 2, \ldots, n. \quad (25.9)$$

The approximation of integrals in (25.4), (25.7) by the quadrature formulas of rectangles with n nodes $\eta = \eta_m = \pi(m - 1/2)/n$, $m = 1, 2, \ldots, n$, leads, with due account of (25.6), (25.8), (25.9) to the following system of linear algebraic equations for the approximate values u_μ^k of the function u at the points P_μ^k:

$$u_\mu^k = (\mu - 1) \frac{\ln |f_{3-\mu}(z_\mu^k)|}{\ln r_{3-\mu,0}} + \frac{\pi}{n} \sum_{m=1}^{n} u_{3-\mu}^m R_{3-\mu}^{km}, \quad (25.10)$$

$$1 \le \mu \le 2, \qquad 1 \le k \le n,$$

where

$$R_\tau^{km} = R_{\tau 0}^{km} - \frac{1}{n}\sum_{q=1}^{n} R_{\tau 0}^{kq} + (2 - \tau + (-1)^\tau \ln|f_\tau(z_{3-\tau}^k)|/\ln r_{\tau 0})/\pi,$$

$$R_{\tau 0}^{km} = R_\tau(z_{3-\tau}^k, \pi(m - 1/2)/n).$$

Let $u_{\nu\mu}^k$, $1 \le \mu \le 2$, $1 \le k \le n$, be the approximate solution of system (25.10) obtained after the νth Seidel iteration at the initial approximation $u_{02}^k = 1$, $1 \le k \le n$. The approximate solution $u_{\mu n}^\nu(z)$ of problem (16.6) on the closed basic block $\overline{T}_\mu^*$, $\mu = 1, 2$, and its conjugate harmonic function $v_{\mu n}^\nu(z)$ have the form

$$u_{\mu n}^\nu(z) = \quad a_{\mu 0}^\nu + (a_{\mu 1}^\nu - a_{\mu 0}^\nu)\frac{\ln|f_\mu(z)|}{\ln r_{\mu 0}}$$

$$+ \frac{\pi}{n}\sum_{k=1}^{n}(u_{\nu\mu}^k - a_\mu^\nu)R_\mu\left(z, \pi\frac{k - 1/2}{n}\right), \qquad (25.11)$$

$$v_{\mu n}^\nu(z) = \quad (a_{\mu 1}^\nu - a_{\mu 0}^\nu)\frac{\arg f_\mu(z)}{\ln r_{\mu 0}}$$

$$+ \frac{\pi}{n}\sum_{k=1}^{n}(u_{\nu\mu}^k - a_\mu^\nu)I_\mu\left(z, \pi\frac{k - 1/2}{n}\right), \qquad (25.12)$$

where R_μ is kernel (25.5), I_μ is its conjugate kernel (obtained by a formal substitution of I_μ for the symbol R_μ on the left-hand side of (25.5) and of the kernel I, defined in (25.3), for the kernel R on the right-hand side), $a_{20}^\nu = 0$, $a_{11}^\nu = 1$,

$$a_\mu^\nu = a_{\mu,\mu-1}^\nu = \frac{1}{n}\sum_{m=1}^{n} u_{\nu\mu}^m.$$

When solving system (25.10), in the process of iteration we calculate the control quantity

$$\varepsilon_{n\nu}^0 = \max_{1 \le p \le 15}|u_{1n}^\nu(z_p) - u_{2n}^\nu(z_p)|,$$

where $z_p = f_1^{-1}(r_{30}\exp\{i\pi(p - 1/2)/15\}) \in \gamma$. The iteration process is terminated at the minimum $\nu \ge 2$ for which condition (15.13) is fulfilled.

In accordance with (25.12), the approximate value ϑ_n of quantity (16.8) can be found from the formula

$$\vartheta_n = 2\Big(\sum_{\mu=1}^{2} \ln r_{\mu 0} / (a_{\mu 1}^{\nu} - a_{\mu 0}^{\nu}) \Big)^{-1},$$

and the approximate values ρ_{1n} and M_n of quantities (16.9) by formulas (17.15).

For the given ϑ_n the arithmetic mean of the increments of the functions $\vartheta_n v_{\mu n}^{\nu}(z)$, $\mu = 1, 2$, in the clockwise traverse of the circle γ is equal to -2π.

The approximate conformal mapping of the domain Ω onto the ring $\rho_{1n} < |\zeta| < 1$, close to the ring $\rho_1 < |\zeta| < 1$ (see Sec. 16), is defined, on the half-open basic blocks $T_{\mu}^* \cap \Omega$, $\mu = 1, 2$, with the aid of the functions

$$\zeta = \zeta_{\mu n}(z) = \exp\{\vartheta_n(u_{\mu n}^{\nu}(z) + iv_{\mu n}^{\nu}(z))\}.$$

In fact, these functions are defined on the closed basic blocks $\overline{T}_{\mu}^*$, $\mu = 1, 2$, and, in particular, on the circle $\gamma = \overline{T}_1^* \cap \overline{T}_2^*$, where they assume closed values.

Table 25.1 presents the results of computations, according to our algorithm, when the domain Ω with a hole of radius $a = 1.5$ was mapped onto a ring. We chose the computation parameters $b = 10$, $\alpha = 0.95$, $\beta = 0.35$. When computing the series in (19.9), we retained the terms for which $h^{2k-2} > 2^{-48}$. The table includes the following quantities: the values of the discreteness parameter n, the number of iterations ν needed to solve system (25.10), the approximate value M_n of the conformal invariant of the domain being mapped. In addition, as indicator of the accuracy of the result, the table gives the values of the quantity

$$E_n^0 = \varepsilon_{n\nu}^0 |\vartheta_n| \max_{1 \le p \le 15} \exp\{\vartheta_n u_{1n}(z_p)\},$$

which is close to

$$\max_{z \in \gamma} ||\zeta_{1n}(z)| - |\zeta_{2n}(z)||,$$

and the values of E_n^1 approximately equal to the deviation from π of the increment of the argument of the function $\zeta_{\mu n}(z)$, $\mu = 1, 2$, when

Table 25.1

n	ν	M_n	E_n^0	E_n^1
5	3	4.882	$6.9 \cdot 10^{-3}$	$5.7 \cdot 10^{-4}$
10	4	4.880 79	$1.1 \cdot 10^{-4}$	$6.5 \cdot 10^{-6}$
15	5	4.880 773	$2.6 \cdot 10^{-6}$	$1.6 \cdot 10^{-7}$
20	6	4.880 772 76	$2.9 \cdot 10^{-8}$	$4.6 \cdot 10^{-9}$
25	7	4.880 772 745	$4.8 \cdot 10^{-10}$	$1.5 \cdot 10^{-10}$
30	7	4.880 772 745	$3.7 \cdot 10^{-11}$	$4.5 \cdot 10^{-12}$

the upper or the lower half of the circle γ is traversed counterclockwise.

Relatively small values of ν speak of a rapid convergence of iterations. The observed increase in the number of coincident decimal digits in the value M_n with a growth of n and also the behavior of the quantities E_n^0 and E_n^1 substantiate the exponential convergence of the block method.

When the domain Ω with a hole of radius $a \to +0$ is conformally mapped onto the ring $\rho_1 < |\zeta| < 1$, we observe the asymptotic behavior

$$\rho_1 \sim \delta(a), \qquad a \to +0,$$

where $\delta(a) = a/8$. The results obtained for a number of decreasing values of a for $b = 8$, $\alpha = 0.5$, $\beta = 0.2$, $n = 20$, are given in Table 25.2. Here $\sigma_n = \rho_{1n}/\delta(a)$, where ρ_{1n} is the approximate value obtained for ρ_1. These results demonstrate the efficiency of the approximate block method of conformal mappings in the case when an infinite domain

Table 25.2

a	σ_n	ν	E_n^0	E_n^1
$0.8 \cdot 10^0$	1.023	11	$2.8 \cdot 10^{-7}$	$2.3 \cdot 10^{-9}$
$0.8 \cdot 10^{-1}$	1.000 23	12	$2.0 \cdot 10^{-11}$	$1.3 \cdot 10^{-11}$
$0.8 \cdot 10^{-2}$	1.000 002 3	12	$6.4 \cdot 10^{-12}$	$7.4 \cdot 10^{-12}$
$0.8 \cdot 10^{-3}$	1.000 000 023	12	$6.1 \cdot 10^{-12}$	$2.0 \cdot 10^{-11}$
$0.8 \cdot 10^{-4}$	1.000 000 000	11	$5.1 \cdot 10^{-12}$	$1.9 \cdot 10^{-11}$

being mapped has a fine structure, namely, a small hole.

It takes several seconds to compute one variant on the BESM-6 computer.

Chapter 4

Approximate Conformal Mapping of Domains with a Periodic Structure by the Block Method

26. Mapping a Domain of the Type of Half-Plane with a Periodic Structure onto a Half-Plane

We first study the construction of a conformal mapping of a domain of the type of half-plane with a boundary defined by a periodic circular polygonal line onto a half-plane. Then, for a special case, we give an algorithm of an approximate conformal mapping of a domain of the type being studied onto a half-plane with the aid of two pairs of special blocks. The mapping is defined by elementary functions. We give the results of computations for two variants of the periodic boundary of the domain substantiating the exponential character of convergence of the block method. In conclusion we give a brief description of the algorithm of an approximate conformal mapping of a domain of the type in question in the general case.

26.1. Construction of a Conformal Mapping of a Domain of the Type of Half-Plane with a Periodic Boundary onto a Half-Plane

A domain of the type of half-plane is a simply-connected domain which contains a half-plane and lies in some half-plane. Let us con-

sider a domain Ω of the type of half-plane lying in the half-plane
Re $z > 0$, $z = x + iy$. We assume that the boundary Γ of the domain Ω is a circular polygonal line periodic in the direction of the
imaginary axis. On every period equal to 2π with respect to y this
polygonal line has a finite number of sides which are line segments or
circular arcs. The sides may overlap, in particular, forming (periodic)
polygonal cuts on the edge of the domain Ω.

Let $\sigma_k \subset \Omega$, $k = 0, \pm 1, \pm 2, \ldots$, be an open ray on the straight
line Im $z = 2k\pi$, with its endpoint A_k lying on Γ. If the endpoint
of the ray σ_k falls in a multiple point of the boundary Γ, then we
understand A_k to be a boundary point attainable from Ω along the
ray σ_k. We denote by γ_k the continuous part of the boundary Γ
between the points A_k and A_{k-1}. When γ_k is traversed from the
point A_k to A_{k-1}, the domain Ω is locally on the left and the points
A_k and A_{k-1} do not belong to γ_k. Furthermore, let $x = x_0 \geq 0$ be a
coordinate of the point A_0 of the real axis.

We set $\gamma'_k = \gamma_k \cup A_k$ and introduce complex variables ζ and w_j,
$j = 1, 2$. Let G^1 be a finite simply-connected domain in the complex
plane w_1 whose boundary Γ^1 is the image of γ'_0 under the mapping

$$w_1 = \exp\{-z\}, \tag{26.1}$$

and A_0^1 be the image of the point A_0 under this mapping. The point
$w_1 = 0$ obviously lies in G^1 and the boundary Γ^1 of the domain G^1
is piecewise-analytic.

Suppose that Δ^1 is Laplace's operator acting in the complex plane
w_1. We consider a function

$$w_2 = f(w_1) \equiv w_1 \exp\{g(w_1)\}, \tag{26.2}$$

where

$$g(w_1) = \chi(w_1) + i\omega(w_1),$$

χ is the solution of Dirichlet's problem

$$\Delta^1 \chi = 0 \quad \text{on} \quad G^1, \qquad \chi = -\ln|w_1| \quad \text{on} \quad \Gamma^1,$$

and ω is a harmonic function conjugate to χ which vanishes at the
point A_0^1.

The first derivatives of the function χ are continuous up to the boundary Γ^1 of the domain G^1, except for some vertices of its angles. In the vicinity of the indicated vertices the derivatives grow, in their order, not faster than the distance to the vertex in the power $-1/2$. This follows, for example, from the results of [50]. Since the behavior of the derivatives of the conjugate function ω is similar, this function is, like χ, continuous up to the boundary of the domain G^1, the vertices of all angles inclusive.

Function (26.2) conformally maps the domain G^1 onto a disk $|w_2| < 1$ (see [23, item 43]). Under this mapping the point $w_1 = 0$ is carried over to the point $w_2 = 0$ and the point $w_2 = 1$ is the image of the boundary point A_0^1. By what was said above, function (26.2) is continuous up to the boundary Γ^1 of the domain G^1. As follows from the results of [17], here the continuous traverse of the boundary Γ^1 in the positive direction is associated with a continuous counterclockwise traverse of the boundary of the disk $|w_2| < 1$.

On the domain Ω we define a function

$$\zeta = -\ln f(\exp\{-z\}), \tag{26.3}$$

where the imaginary part of the logarithm, i.e., $\arg f(\exp\{-z\})$ is a continuous branch with respect to z, and

$$\lim_{x \to x_0 + 0} \arg f(\exp\{-x\}) = 0.$$

By virtue of (26.1) and the properties of the function f indicated above, this function is analytic in the domain Ω and continuous up to its boundary Γ. Here function (26.3) maps Ω one-to-one onto the half-plane Re $\zeta > 0$ and the continuous traverse of the boundary Γ is associated with a continuous traverse of the boundary of the half-plane Re $\zeta > 0$ in the same direction. Thus function (26.3) conformally maps one-to-one the domain Ω of the type of half-plane onto the haf-plane Re $\zeta > 0$, the point $z = \infty$ passing into the point $\zeta = \infty$ and the point $\zeta = 0$ serving as the image of the point A_0. With due account of (26.2), we write function (26.3) as

$$\zeta = z + t(z), \tag{26.4}$$

where

$$t(z) = -g(\exp\{-z\}). \qquad (26.5)$$

When the boundary points $\zeta = 0$ and $\zeta = \infty$ are stationary, the conformal automorphisms of the half-plane $\text{Re}\,\zeta > 0$ are described by the linear function $c\zeta$, where $c > 0$ is real. Therefore, the conformal mapping of the domain Ω onto the half-plane $\text{Re}\,\zeta > 0$, representable in form (26.4), where $t(z)$ is a bounded analytic function, is unique when two pairs of boundary points are in the correspondence being considered. Also note that according to (26.5) the function $t(z)$ has a period 2π in the direction of the imaginary axis.

26.2.   Constructing Blocks

Suppose a domain Ω of the type of half-plane shown in Fig. 35 is given in the complex plane z. Its boundary Γ, 2π-periodic in the direction of the imaginary axis, consists, on every period, of two circular arcs, that form a cog, and a segment of the imaginary axis connecting two adjacent cogs. The cog whose vertex B lies on the real axis is formed by the circular arcs AB and CB orthogonal to the imaginary axis. It is h_0 high and $h_1 + h_2$ wide at the base.

We set $u(z) = \text{Re}\,t(z)$, $v(z) = \text{Im}\,t(z)$. Then function (26.4), that conformally maps the domain Ω onto the half-plane $\text{Re}\,\zeta > 0$, assumes the form

$$\zeta = z + u(z) + iv(z), \qquad (26.6)$$

where u is the only bounded solution of the boundary-value problem

$$\Delta u = 0 \quad \text{on} \quad \Omega, \qquad u = -x \quad \text{on} \quad \Gamma, \qquad (26.7)$$

and v is a harmonic function conjugate to u, with $v(h_0) = 0$.

Since the functions u and v are 2π-periodic in the direction of the imaginary axis, it suffices to find them on the closed subdomain $\overline{\Omega}_0$, where

$$\Omega_0 = \Omega \cap \{z : -\pi - \delta < y < \pi - \delta\}, \qquad (26.8)$$

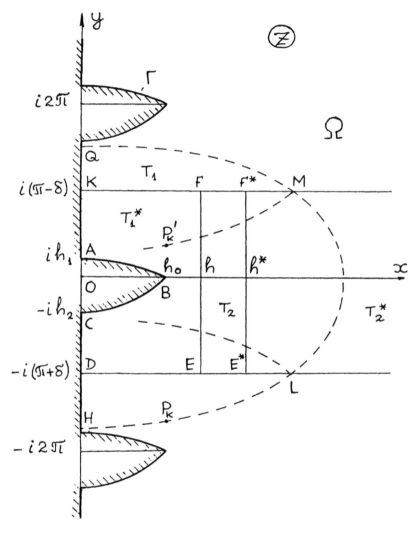

Figure 35

δ is a number. Below we construct approximations for the functions u and v with the aid of elementary harmonic functions on two closed subdomains, called blocks, whose union forms a set $\overline{\Omega}_0$.

Let us carry out some constructions. We specify real parameters $h_0 > 0$, $h_1 \geq 0$, $h_2 > 0$, h, h^*, δ, $\varkappa$, $r_0 > 1 + |\varkappa|$, with $h^* > h > h_0$, $h_1 + h_2 < 2\pi$ (see, in particular, Fig. 35). Later we shall indirectly touch upon the other restrictions imposed on the parameters. Let

$$\alpha_q = \tan^{-1}(h_q/h_0), \qquad q = 1, 2,$$

$$\alpha_0 = \alpha_1 - \alpha_2, \qquad \alpha = \alpha_1 + \alpha_2,$$

$$w_3(z) = ((z - h_0)e^{i\alpha_0}/(z + h_0))^{\pi/(\pi-\alpha)},$$

$$w_4(\tau) = i(((1 + \tau)/(1 - \tau))^2 - 1)^{1/2}.$$

The function

$$w = w(z) \equiv w_4(w_3(z)) - \varkappa \tag{26.9}$$

conformally maps the half-plane Re $z > 0$, which has a single cut, namely, a cog ABC (Fig. 35), onto the half-plane Im $w > 0$. Here the points A, B, C are carried over to the points $w = -1 - \varkappa$, $w = -\varkappa$, $w = 1 - \varkappa$, respectively. We denote by $z(w)$ an elementary function that is the inverse of $w(z)$.

We set

$$T_1 = \{z : |w(z)| < r_0, \text{ Im } w(z) > 0\},$$

$$T_2 = \{z : x > h, -\pi - \delta < y < \pi - \delta\}.$$

We call T_2 an extended block of the type of half-strip and T_1 an extended block of the type of half-disk and introduce basic blocks

$$T_1^* = \Omega \cap \{z : x < h^*, -\pi - \delta < y < \pi - \delta\},$$

$$T_2^* = \{z : x > h^*, -\pi - \delta < y < \pi - \delta\}.$$

The extended block T_1 is the inverse image of the half-disk

$$\{w : |w| < r_0, \text{ Im } w > 0\}$$

under the mapping $w = w(z)$. The part of the boundary of this block lying in Ω is shown in Fig. 35 by a dash line $HLMQ$. The basic block T_1^* is a circular polygon $ABCDE^*F^*K$.

We assume that the parameters h_0, h_1, h_2, h^*, δ, $\varkappa$, r_0 take on values for which the arrangement of the blocks is that shown in Fig. 35, i.e.,

$$T_\mu^* \subset T_\mu \subset \Omega_0, \qquad \overline{T}_\mu^* \cap \Omega_0 \subset T_\mu, \qquad \mu = 1, 2,$$
$$\overline{T}_1^* \cup \overline{T}_2^* = \overline{\Omega}_0.$$

26.3.   The Algorithm

Let

$$\frac{R_1}{I_1} (z, \eta) = \sum_{p=0}^{1} (-1)^p \frac{R}{I} \left(\frac{|w(z)|}{r_0}, \arg w(z), (-1)^p \eta \right),$$

$$\frac{R_2}{I_2} (z, \eta) = \frac{R}{I} \left(\exp\{h - x\}, \pi - \delta - y, \eta \right),$$

where $R(r, \theta, \eta)$, $I(r, \theta, \eta)$ are conjugate kernels (3.14), (8.13).

We have the following representations for the functions u and v:

$$\frac{u}{v} (z) = \frac{-x}{C_1 - y} + \int_0^\pi (u(z(r_0 e^{i\eta})) + \operatorname{Re} z(r_0 e^{i\eta})) \frac{R_1}{I_1} (z, \eta) \, d\eta$$

$$(26.10)$$

on $T_1 \cup \overline{T}_1^*$,

$$\frac{u}{v} (z) = \frac{0}{C_2} + \int_0^{2\pi} (u(h + i(\pi - \delta - \eta))) \frac{R_2}{I_2} (z, \eta) \, d\eta \qquad (26.11)$$

on $T_2 \cup \overline{T}_2^*$, where C_1, C_2 are constants. Relations (26.10), (26.11) can be proved in the same way as Theorem 3.1.

We choose a natural number $n > 20$ and specify points P_k, $k = 1, 2, \ldots, 2n - 20$, with complex coordinates

$$z_k = x_k + iy_k = \begin{cases} z(\xi_k), & 1 \le k \le n, \\ h + i\left(\pi - \delta - \dfrac{\pi(2k - 2n - 1)}{n - 20} \right), & k > n, \end{cases}$$

where $\xi_k = r_0 \exp\{i\theta_k\}$, $\theta_k = \pi(2k-1)/2n$. For $1 \le k \le n$ the points P_k lie on the part of the boundary of the block T_1 which is in Ω (the curve $HLMQ$ in Fig. 35), and for $n < k \le 2n - 20$ on the line segment EF belonging to the boundary of the block T_2. With every point P_k we associate a point P'_k with a complex coordinate

$$z'_k = x'_k + iy'_k = x_k + i(2\pi\{(y_k + \pi + \delta)/2\pi\} - \pi - \delta),$$

where $\{\ \}$ is the sign of the fractional part, and set $\tau_k = 1$ for $x_k \le h^*$, $\tau_k = 0$ for $x_k > h^*$.

We have

$$P'_k \in \begin{cases} \overline{T^*_1}, & \tau_k = 1, \\ \overline{T^*_2}, & \tau_k = 0, \end{cases} \tag{26.12}$$

i.e., the points P'_k, $1 \le k \le 2n - 20$, lie on the closed basic blocks and, in addition, by virtue of the periodicity of the function u

$$u(z_k) = u(z'_k), \qquad 1 \le k \le 2n - 20. \tag{26.13}$$

Approximating the integral in (26.10) at $z = z'_k$ for the values of k for which $\tau_k = 1$, and the integral in (26.11) at $z = z'_k$, $\tau_k = 0$ by the quadrature formula of rectangles with n and $n - 20$ nodes, respectively, we arrive, with due regard for (26.12) and (26.13) at the following system of linear algebraic equations for the approximate values u^k of the function u at the points P_k:

$$u^k = \begin{cases} -x_k + \dfrac{\pi}{n} \displaystyle\sum_{m=1}^{n}(u^m + x_m)R_1(z'_k, \theta_m), & \tau_k = 1, \\[4mm] \dfrac{2\pi}{n-20} \displaystyle\sum_{m=1}^{n-20} u^{n+m} R_2\Big(z'_k, \dfrac{\pi(2m-1)}{n-20}\Big), & \tau_k = 0, \end{cases} \tag{26.14}$$

$$1 \le k \le 2n - 20,$$

where $\theta_m = \pi(2m-1)/2n$. System (26.14) can be solved by means of Seidel's iterations at the zero initial approximation. In each iteration we first carry out computations for successive values of k corresponding to $\tau_k = 1$ and then for the remaining k. We denote by u^k_ν, $1 \le k \le 2n - 20$, the approximate solution of system (26.14), obtained after the νth iteration.

Replacing the integrals in (26.10), (26.11) by the quadrature formulas of rectangles and the unknown values of the function u at the
nodes by their approximate values u_ν^k, $1 \le k \le 2n - 20$, we get the
following approximations $u_{\mu n}^\nu$, $v_{\mu n}^\nu$ for the functions u and v on the
closed basic blocks $\overline{T}_\mu$, $\mu = 1, 2$,

$$
\begin{matrix} u_{1n}^\nu \\ v_{1n}^\nu \end{matrix} (z) = \frac{-x}{C_1 - y} + \frac{\pi}{n} \sum_{k=1}^{n} (u_\nu^k + x_k) \frac{R_1}{I_1} (z, \theta_k),
$$

$$
\begin{matrix} u_{2n}^\nu \\ v_{2n}^\nu \end{matrix} (z) = \begin{matrix} 0 \\ C_2 \end{matrix} + \frac{2\pi}{n - 20} \sum_{m=1}^{n-20} u_\nu^{n+m} \frac{R_2}{I_2} \left(z, \frac{\pi(2m - 1)}{n - 20} \right).
$$

These functions are elementary. The constant C_1 can be found
from the condition $v_{1n}^\nu(h_0) = 0$, and C_2 from the equality $v_{2n}^\nu(h^*) = v_{1n}(h^*)$.

We set $t_{\mu n}^\nu(z) = u_{\mu n}^\nu(z) + i v_{\mu n}^\nu(z)$,

$$
\zeta_{\mu n}^\nu(z) = z + t_{\mu n}^\nu(z) \quad \text{on} \quad \overline{T}_\mu^*, \qquad \mu = 1, 2. \tag{26.15}
$$

Formulas (26.15) define approximately the desired conformal mapping on subdomain (26.8). Mapping (26.15) can be extended to the
entire domain Ω by means of a 2π-periodic continuation of the function $t_{\mu n}^\nu$, $\mu = 1, 2$ in the direction of the imaginary axis. The function
ζ_{2n}^ν is continued analytically whereas the continued function ζ_{1n}^ν has
discontinuities of the first kind (to within the error) on the segments
lying on the straight lines $y = \pi - \delta + 2\pi q$, $q = 0, \pm 1, \pm 2, \ldots$.

26.4.   Practical Results

When solving system (26.14), in the process of iteration we calculate the control quantity

$$
\varepsilon_{n\nu}^0 = \max_{1 \le p \le 15} |u_{1n}^\nu(z_p^*) - u_{2n}^\nu(z_p^*)|,
$$

where $z_p^* = h^* + i(\pi - \delta - \pi(2p - 1)/15)$, characterizing the deviation
of the function u_{1n}^ν and u_{2n}^ν from each other on the line segment E^*F^*
(Fig. 35), coinciding with $\overline{T}_1^* \cap \overline{T}_2^*$. Because of their rapid convergence,
the iterations terminate at the minimum $\nu \ge 2$, for which condition
(15.13) is fulfilled.

After terminating the iterations we find the second control quantity

$$\varepsilon^1_{n\nu} = \max_{1 \le q \le 15} |v^\nu_{1n}(z^0_q) - v^\nu_{1n}(z^1_q)|,$$

where $z^m_q = qh^*/15 + i(\pi - \delta - 2\pi m)$, characterizing a jump of the function v^ν_{1n} (approximating the function v) on the line segments DE^* and KF^* in its 2π-periodic continuation from the block T^*_1 in the direction of the imaginary axis.

Here are the results of the computations carried out for two variants.

(1) Let $h_0 = 5$, $h_1 = 0.5$, $h_2 = 1$, $h = 5.9$, $h^* = 6.35$, $\delta = 0.04$ (see Fig. 35), $\varkappa = -0.04$, $r_0 = 1.37$. The cog ABC being considered has no symmetry and its height h_0 is commensurable with the period 2π. The vertex angle of the cog on the side of the domain Ω is equal to $\approx 1.81\pi$, i.e., exceeds π. Therefore (see, for instance, [50]) even the first derivatives of the functions u and v are unbounded in the neighborhood of the vertices of the cogs. However, since the distance from the points P_k and P'_k, $1 \le k \le 2n - 20$, to the indicated vertices exceeds some positive constant, independent of n, the existence of singularities of the derivatives of the function u at the vertices does not affect the exponential character of the convergence of the block method. At the bases of the cogs the domain Ω has angles equal to $\pi/2$ formed by the line segments and the circular arcs. In the neighborhood of the vertices of these angles, including the vertices themselves, the functions u and v admit of an analytic continuation across the boundary of the domain Ω. Therefore the points P_k and P'_k can lie arbitrarily close to the bases of the cogs. The results of the computations carried out for this variant are presented in Table 26.1.

Here n is the discreteness parameter, ν is the number of iterations needed to solve system (26.14), $u^\nu_{2n}(\infty)$ is the asymptotic value of the function $u^\nu_{2n}(z)$ in the positive direction of the real axis, $\varepsilon^0_{n\nu}$, $\varepsilon^1_{n\nu}$ are indicators of accuracy.

(2) We set $h_0 = 1$, $h_1 = h_2 = 2 > h_0$. Then the circular arcs AB and CB, orthogonal to the imaginary axis, form, instead of an acute cog (Fig. 35), a symmetric projection ABC (Fig. 36), indented at the middle which is repeated on the boundary of the domain Ω

Table 26.1

n	ν	$u^{\nu}_{2n}(\infty)$	$\varepsilon^0_{n\nu}$	$\varepsilon^1_{n\nu}$
30	8	-3.75	$8.7 \cdot 10^{-2}$	$2.7 \cdot 10^{-2}$
45	19	$-3.76\ 334$	$1.1 \cdot 10^{-4}$	$1.6 \cdot 10^{-5}$
60	27	$-3.76\ 335\ 09$	$1.2 \cdot 10^{-7}$	$9.8 \cdot 10^{-8}$
75	34	$-3.76\ 335\ 1024$	$2.8 \cdot 10^{-10}$	$7.8 \cdot 10^{-10}$
80	35	$-3.76\ 335\ 1024$	$1.6 \cdot 10^{-10}$	$6.9 \cdot 10^{-10}$

of the type of half-plane with period 2π. In this case the algorithm
of construction of an approximate conformal mapping additionally
takes into account the evenness of the function u relative to the real
axis. Hence, when n is even and $\delta = \varkappa = 0$, the number of unknowns
in system (26.14) decreases by half. The results obtained for $h = 2$,
$h^* = 2.4$, $\delta = \varkappa = 0$, $r_0 = 1.4$ are given in Table 26.2.

Tables 26.1 and 26.2 show a distinct exponential convergence of
the block method. The average time of computation on the BESM-6
computer of one variant for n being considered is 10 seconds.

26.5.   The General Case

Let Ω be a domain of the type of half-plane described in 26.1.
We denote by Ω_0 its subdomain bounded by rays σ_0 and σ_1, their
endpoints inclusive, and the circular polygonal line γ_1.

We specify the numbers h^* and h such that $h^* > h > 0$, with
$\overline{T}_0 \cap \Gamma = \varnothing$, where Γ is the boundary of the domain Ω,

$$T_0 = \{z : x > h, 0 < y < 2\pi\}$$

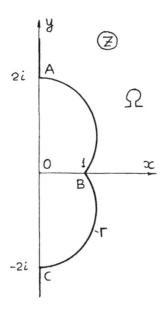

Figure 36

Table 26.2

$n/2$	ν	$u_{2n}^{\nu}(\infty)$	$\varepsilon_{n\nu}^{0}$	$\varepsilon_{n\nu}^{1}$
15	6	-0.928	$3.8 \cdot 10^{-2}$	$1.8 \cdot 10^{-2}$
25	14	$-0.932\ 901$	$1.3 \cdot 10^{-5}$	$8.1 \cdot 10^{-6}$
35	25	$-0.932\ 902\ 3468$	$5.6 \cdot 10^{-9}$	$3.5 \cdot 10^{-9}$
45	26	$-0.932\ 902\ 3468$	$6.8 \cdot 10^{-11}$	$9.3 \cdot 10^{-11}$

is an extended block-half-strip. Also suppose that

$$T_0^* = \{z : x > h^*, 0 < y < 2\pi\}$$

is a basic block-half-strip, and

$$\Omega^* = \Omega_0 \setminus \overline{T_0^*}$$

is a finite circular polygon. It is obvious that $\overline{\Omega}_0 = \overline{T_0^*} \cup \overline{\Omega}_0^*$, with $\overline{T_0^*} \cap \overline{\Omega}^*$ being a segment of a line $x = h^*$.

Using the methods presented in Sec. 2 (if γ_1 is a polygonal line that does not contain any circular arcs) or in [67] (for the general case), we define a finite number of bounded basic and extended blocks T_μ^* and $T_\mu \subset \Omega$, $(\overline{T}_\mu^* \cap \Omega) \subset T_\mu$, $\mu = 1, 2, \ldots, M$, from the standard collection. In a special case when γ_1 does not contain any circular arcs, we use blocks of only three types, namely, sectors of disks, half-disks, and disks (see Sec. 2). For the general case, the description of all standard blocks is given in [67]. The same requirements are imposed on the arrangement of every pair of blocks T_μ and T_μ^* in Ω as in Sec. 2 and [67] (respectively). The objective here is not to get a covering for the whole domain Ω but only to satisfy the condition

$$\bigcup_{\mu=1}^{M} \overline{T}_\mu^* \supset \overline{\Omega}^*. \tag{26.16}$$

To approximate function (26.6) which defines the required conformal mapping of the domain Ω, we first seek an approximate solution of the boundary-value problem (26.7) on the closed basic blocks $\overline{T}_\mu^*$, $\mu = 0, 1, \ldots, M$. With this aim in view, we specify $n_0 = n$ points P_{0k}, $k = 1, 2, \ldots, n_0$, with a complex coordinate $z_{0k} = h + i2\pi(1 - (k - 1/2)/n_0)$ on the boundary of the extended block-half-strip T_0. On the boundaries of the other extended blocks T_μ, $\mu = 1, 2, \ldots, M$, we choose $n_\mu = O(n)$ points $P_{\mu k}$, $k = 1, 2, \ldots, n_\mu$, by the technique indicated in Sec. 2 or in [67]. We associate every point $P_{\mu k}$, which has a complex coordinate $z_{\mu k} = x_{\mu k} + iy_{\mu k}$, $0 \le \mu \le M$, $1 \le k \le n_\mu$, with a point $P_{\mu k}'$ with a complex coordinate $z_{\mu k}' = x_{\mu k}' + iy_{\mu k}'$. Here we set $x_{\mu k}' = x_{\mu k}$, and choose $y_{\mu k}'$ such that the following two conditions are fulfilled:

$$\{(y_{\mu k}' - y_{\mu k})/2\pi\} = 0,$$

where { } is the sign of the fractional part, and

$$P'_{\mu k} \in \overline{\Omega}^* \cup \overline{T}_0^*, \qquad 0 \leq \mu \leq M, \qquad 1 \leq k \leq n_\mu. \qquad (26.17)$$

By virtue of the periodicity of the function u with respect to y, the equalities

$$u(z_{\mu k}) = u(z'_{\mu k}), \qquad 0 \leq \mu \leq M, \qquad 1 \leq k \leq n_\mu, \qquad (26.18)$$

hold true.

Then as we did in Sec. 4 (in the case when a polygonal line does not contain circular arcs) and as it is done in [67] (in the general case), with due account of (26.16)–(26.18) and of the boundary condition in (26.7), we set up a system of linear algebraic equations for the approximate values u_μ^k of the function u at the points $P_{\mu k}$, $0 \leq \mu \leq M$, $1 \leq k \leq n_\mu$, by approximating the integral representations of the function u on blocks by quadrature formulas. For a sufficiently large n this system is uniquely solvable.

A repeated approximation of the integral representations of the function u by the quadrature formulas of rectangles with a substitution of the approximate values u_μ^k of the function u at the nodes leads to the approximation of the function u on closed basic blocks $\overline{T}_\mu^*$, $0 \leq \mu \leq M$, by harmonic functions. A formal replacement of the obtained harmonic functions by their conjugates with an addition of arbitrary constants yields an approximation of the function v separately on every closed basic block $\overline{T}_\mu^*$, $0 \leq \mu \leq M$. The concordance of the arbitrary constants for all blocks is performed according to the methods presented in Sec. 9 with due regard for the condition $v(x_0) = 0$. The approximate expression for the required conformal mapping (26.6) obtained on blocks has a form similar to (26.15). In particular, the obtained approximate conformal mapping of the subdomain

$$\Omega_0 \subset \bigcup_{\mu=0}^{M} \overline{T}_\mu^*$$

extends to the whole domain Ω with due account of (26.6) and the 2π-periodicity of the functions u and v in the direction of the imaginary axis.

Remark 26.1. The general algorithm for constructing an approximate conformal mapping of a domain of the type of half-plane, bounded by an arbitrary periodic circular polygonal line, onto a half-plane, that we have briefly described, does not require the use of the apparatus of analytic continuation (all constructions are within the bounds of the given domain) or any replacement of the complex variable (leading to special blocks) or a symmetry (provided that it exists). In special cases this algorithm can be improved. For example, a direct use of this algorithm for mapping the domain described in 26.2 requires the specification of at least five pairs of standard blocks. However, the introduction of a local change of the complex variable (26.9), which can be used to construct the special block T_1 of the type of a half-disk, made it possible to reduce the number of pairs of blocks to two in the cases considered in 26.4. Below, in Sec. 27, we use the analytic continuation of the required function when mapping a special domain of the type of a strip onto a strip and thus reduce the number of pairs of the blocks to one pair.

27.   Mapping a Domain of the Type of Strip with a Periodic Structure onto a Strip

This section is constructed in the same way as the preceding section.

27.1.   Contruction of a Conformal Mapping of a Domain of the Type of Strip with a Periodic Boundary onto a Strip

A domain of the type of strip is an infinite simply-connected domain which lies in a strip and is such that any sraight line that cuts the strip also cuts the domain. Let us consider a domain Ω of the type of strip lying in the strip $0 < \mathrm{Re}\, z < 1$. We assume that the boundary of the domain Ω is formed by two circular polygonal lines Γ^0 and Γ^1, that are periodic in the direction of the imaginary axis and consist, on the common period l of a finite number of line segments and circular arcs, the distance from Γ^0 to the imaginary axis being smaller than that from Γ^1.

Let $\sigma_k \subset \Omega$ be an open interval of the straight line $y = kl$, $k = 0, \pm 1, \pm 2, \ldots$, whose endpoints A_k^j lie on Γ^j, $j = 0, 1$, respectively. Here the points A_k^j, $j = 0, 1$, $k = 0, \pm 1, \pm 2, \ldots$, in their collection, lie on two parallel straight lines and A_0^0 is the one of the points Γ^0, lying on the real axis that is the most distant from $z = 0$.

We assume that the function

$$\zeta = u + iv \tag{27.1}$$

conformally maps the domain Ω onto the strip $0 < \text{Re } \zeta < 1$. Under this mapping the point $\zeta = 0$ corresponds to the point A_0^0 and the polygonal lines Γ^j are carried over to the straight lines $\text{Re } \zeta = j$, $j = 0, 1$, respectively. This function is unique and its real part u is a bounded solution of the boundary-value problem

$$\Delta u = 0 \quad \text{on} \quad \Omega, \qquad u = j \quad \text{on} \quad \Gamma^j, \qquad j = 0, 1, \tag{27.2}$$

which obviously satisfies the identity

$$u(z + il) \equiv u(z) \quad \text{on} \quad \Omega. \tag{27.3}$$

The function v is a harmonic function conjugate to u and vanishes at the point A_0^0, and, by virtue of (27.3), the identity

$$v(z + il) - v(z) \equiv L \quad \text{on} \quad \Omega, \tag{27.4}$$

where $L > 0$ is a constant to be defined, holds true.

27.2.   Constructing Blocks

Let us consider a domain Ω of the type of strip shown in Fig. 37. It is bounded by a straight line $\Gamma^1 = \{z : x = 1\}$ and a circular polygonal line Γ^0 formed by circles, of diameter $d < 1$, arranged periodically in the direction of the imaginary axis and touching the imaginary axis, and line segments connecting the points of tangency of adjacent circles. The distance between the centers of adjacent circles $l > d$. The points of tangency $z = ikl$, $k = 0, \pm 1, \pm 2, \ldots$, are

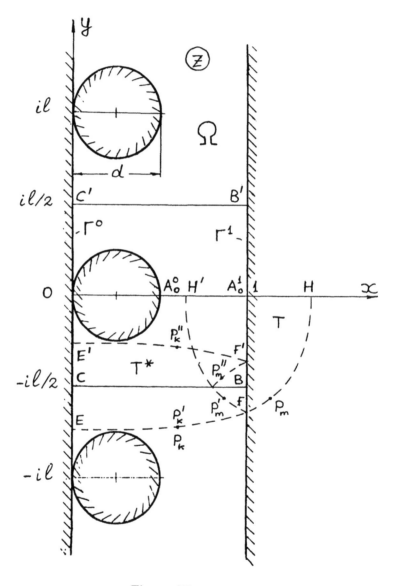

Figure 37

actually multiple points for Γ^0. The vertices of two zero angles of the domain Ω lie at each of the indicated points.

The elementary function

$$w = w(z) \equiv -\cot (i\pi d/2z) \qquad (27.5)$$

conformally maps a half-plane with a single round cut at the edge, i.e.,

$$\Pi = \{z : \operatorname{Re} z > 0, |z - d/2| > d/2\},$$

onto a half-plane $\operatorname{Im} w > 0$. Under this mapping the boundary point $z = d$ is carried over to the point $w = 0$, the part of the boundary of the domain Π formed by the circle is rectified into an interval $[-1,1]$ of the real axis $\operatorname{Im} w = 0$, with the endpoints of the indicated intervals corresponding to the multiple point $z = 0$.

We denote by $z(w)$ the elementary function that is the inverse of (27.5). We specify a number $r_0 > 1$ and call

$$T = \{z : |w(z)| < r_0, \operatorname{Re} w(z) > 0, \operatorname{Im} w(z) > 0\}$$

an extended block of the type of sector. The subdomain

$$T^* = \{z : z \in \Omega, -l/2 < \operatorname{Im} z < 0\}$$

will be a basic block. We assume that d and r_0 are such that the arrangement of the blocks is that shown in Fig. 37, where $EFHA_0^0O$ is the extended block T, and $CBA_0^1A_0^0O$ is the basic block T^*. To be more precise, the condition $T^* \subset T$ is fulfilled, the distance from the basic block T^* to the part EFH of the boundary of the extended block T, shown by a dash line, being positive, and moreover,

$$z'(z) \in \overline{\Omega}, \qquad \forall z \in \overline{T}, \qquad (27.6)$$

where $z'(z) = z = x + iy$ for $x \leq 1$, $z'(z) = 2 - x + iy$ for $x > 1$.

27.3.   The Algorithm

Let

$$\frac{R_0}{I_0} (z, \eta) = \sum_{q=0}^{1} \sum_{p=0}^{1} (-1)^p \frac{R}{I} \left(\frac{|w(z)|}{r_0}, \arg w(z), \varphi_{pq}(\eta) \right),$$

where $\varphi_{pq}(\eta) = (-1)^p(\pi q + (-1)^q \eta)$, R, I are conjugate kernels (3.14), (8.13). We retain the same notations for the analytic continuations of the functions u and v across Γ^1. We can represent these functions on $T \cup \overline{T}^*$ as

$$\begin{matrix} u \\ v \end{matrix}(z) = \int_0^{\pi/2} u(z(r_0 e^{i\eta})) \begin{matrix} R_0 \\ I_0 \end{matrix}(z, \eta)\, d\eta, \qquad (27.7)$$

which can be proved in the same way as Theorem 3.1.

We choose a natural number n and specify points P_k, $1 \le k \le n$, with complex coordinates

$$z_k = x_k + iy_k = z(w_k),$$

where $w_k = r_0 \exp\{i\pi(2k-1)/4n\}$. These points lie on the curve EFH (Fig. 37). With every point P_k we associate a point P_k' with a complex coordinate

$$z_k' = x_k' + iy_k' = \begin{cases} x_k + iy_k, & x_k \le 1, \\ 2 - x_k + iy_k, & x_k > 1, \end{cases}$$

a point $P_k'' \in \overline{T}^*$ with a complex coordinate

$$z_k'' = x_k'' + iy_k'' = \begin{cases} x_k' + iy_k', & y_k' \ge -l/2, \\ x_k' + i(-y_k' - l), & y_k' < -l/2, \end{cases}$$

and also a quantity

$$\tau_k = \begin{cases} 1, & x_k \le 1, \\ -1, & x_k > 1. \end{cases}$$

Since the function u is even with respect to the straight line CB (Fig. 37) and can be analytically continued across the straight line $x = 1$, on which $u = 1$, we have

$$u(z_k) = 1 - \tau_k + \tau_k u(z_k''), \qquad 1 \le k \le n. \qquad (27.8)$$

Approximating the upper integral for $z = z_k''$, $k = 1, 2, \ldots, n$, in (27.7) by the quadrature formula of rectangles and substituting at the nodes P_m the approximate unknown values u^m, $m = 1, 2, \ldots, n$, of the function u for its unknown values, we arrive, with due account of (27.8), at a system of linear algebraic equations

$$u^k = 1 - \tau_k + \tau_k \beta \sum_{m=1}^{n} u^m R_0(z_k'', \theta_m), \qquad (27.9)$$

$$1 \leq k \leq n,$$

where $\beta = \pi/2n$, $\theta_m = \pi(2m-1)/4n$.

Let u_ν^k, $1 \leq k \leq n$, be the approximate solution of this system obtained after the νth Seidel's iteration at the initial approximation $u_0^k = 2k/n$. Again replacing the integral in (27.7) by the quadrature formula of rectangles, substituting the approximate values u_ν^m of the function u at the nodes P_m, we get approximations $u_{n\nu}(z)$ and $v_{n\nu}(z)$ for the functions u and v on the closed block $\overline{T}^*$:

$$\begin{matrix} u_{n\nu} \\ v_{n\nu} \end{matrix} (z) = \beta \sum_{m=1}^{n} u_\nu^m \frac{R_0}{I_0} \left(z, \frac{\pi(2m-1)}{4n} \right).$$

It can be directly verified that the identity

$$v_{n\nu}(x) \equiv 0, \qquad d \leq x \leq 1,$$

holds true for any n and ν, i.e., in particular, the function $v_{n\nu}(z)$ vanished at the point $z = d$. In addition, $u_{n\nu}(d) = 0$.

We set

$$\zeta_{n\nu}(z) = u_{n\nu}(z) + iv_{n\nu}(z) \quad \text{on} \quad \overline{T}^*. \qquad (27.10)$$

The real parts of functions (27.1) and (27.10) are even with respect to the real axis and their imaginary parts are odd. Therefore it is natural to assume function (27.10) to be an approximation of function (27.1) not only on the basic block $BA_0^1A_0^0OC$ but also on the circular polygon $BB'C'OA_0^0OC$ (see Fig. 37).

27.4.   Practical Results

When using iterations to solve system (27.9), we calculate the control quantity

$$\varepsilon_{n\nu}^0 = \max_{1 \leq p \leq 15} |u_{n\nu}(z_p) - 1|,$$

where $z_p = 1 - il(p-1)/28$, characterizing the deviation of the function $u_{n\nu}$ from unity, i.e., from u on the line segment BA_0^1. The iterations are terminated at the minimum $\nu \geq 2$ satisfying condition (15.13).

Since the function u is even with respect to the straight lines $y = kl/2$, $k = 0, \pm 1, \pm 2, \ldots$, it follows that the conjugate function v is constant at the intersection of these straight lines with the domain Ω being considered. To be more precise, for $k = 0, \pm 1, \pm 2, \ldots$, we have

$$v(x + ikl/2) \equiv kL/2, \qquad (1 + (-1)^k)d/2 \leq x \leq 1, \qquad (27.11)$$

where L is the unknown constant from identity (27.4).

Taking into account (27.11), we find the following approximate value for L:

$$L_n = -\frac{2}{15} \sum_{q=1}^{15} v_{n\nu}(z_q'),$$

where $z_q' = (q-1)/14 - il/2$, and the second control quantity

$$\varepsilon_{n\nu}^1 = \max_{1 \leq q \leq 15} |v_{n\nu}(z_q') + L_n/2|,$$

characterizing the deviation of the function $v_{n\nu}$ from a constant on the line segment CB (Fig. 37).

Table 27.1 shows the results of computations carried out for $d = 0.3$, $l = 1.7$, $r_0 = 3.3$ on the BESM-6 computer and Table 27.2 gives the results for $d = 0.6$, $l = 1.2$, $r_0 = 1.253$, which illustrate the efficiency of the exponentially converging block method. Note that a small number of iterations ν are needed for solving system (27.9) with a sufficient accuracy. The average time of computing one variant for the numbers n being considered is approximately 2 seconds.

27.5.   The General Case

Suppose that Ω is an arbitrary domain of the type of strip, described in 27.1, and Ω^* is its subdomain bounded by line segments $A_0^0 \cup \sigma_0 \cup A_0^1$, $A_1^0 \cup \sigma_1 \cup A_1^1$ and circular polygonal lines $A_0^0 A_1^0 \subset \Gamma^0$, $A_0^1 A_1^1 \subset \Gamma^1$. We specify a finite covering of Ω^* by standard blocks

Table 27.1

n	ν	L_n	$\varepsilon_{n\nu}^0$	$\varepsilon_{n\nu}^1$
5	2	1.946 6	$6.0 \cdot 10^{-2}$	$6.1 \cdot 10^{-2}$
20	6	1.946 263 4	$2.1 \cdot 10^{-6}$	$2.0 \cdot 10^{-6}$
35	9	1.946 263 793	$1.1 \cdot 10^{-10}$	$4.3 \cdot 10^{-10}$
45	9	1.946 263 793	$8.1 \cdot 10^{-11}$	$3.1 \cdot 10^{-11}$

Table 27.2

n	ν	L_n	$\varepsilon_{n\nu}^0$	$\varepsilon_{n\nu}^1$
15	3	2.348	$2.2 \cdot 10^{-3}$	$9.3 \cdot 10^{-3}$
35	6	2.349 073	$5.2 \cdot 10^{-6}$	$4.9 \cdot 10^{-6}$
55	9	2.349 072 253	$5.3 \cdot 10^{-9}$	$8.3 \cdot 10^{-10}$
75	10	2.349 072 253	$1.4 \cdot 10^{-10}$	$1.9 \cdot 10^{-11}$

that satisfies a condition of form (26.16) and find an approximate solution of the boundary-value problem (27.2) on blocks by the same methods that we used in 26.5. The only difference is that in this case there is no block-half-strip. We use the approximations of the function u on blocks to construct approximations for v as we did in 26.5. Thus we find the approximation for function (27.1) on $\overline{\Omega}^*$. We can extend the approximate conformal mapping from Ω^* to the entire domain Ω by virtue of properties (27.3), (27.4). Here the increment of the approximation for v between the points A_0^0 and A_1^0 or between A_0^1 and A_1^1 serves as an approximation of the quantity L.

28. Mapping the Exterior of a Lattice of Ellipses onto the Exterior of a Lattice of Plates

In this section the block method is used to perform an approximate conformal mapping of an infinite infinitely-connected domain.

28.1. Scheme of Mapping

Suppose that Ω is an infinitely-connected domain in the complex plane z, $z = x + iy$, bounded by a lattice (family) of ellipses

$$\Gamma_p = \left\{ z : \frac{(x - 2p)^2}{a^2} + \frac{y^2}{b^2} = 1 \right\}, \qquad p = 0, \pm 1, \pm 2, \ldots,$$

where $0 < a \le b$, $a < 1$. Suppose, furthermore, that

$$\Omega_0 = \Omega \cap \{ z : 0 < x < 1 \} \tag{28.1}$$

is a simply-connected subdomain of the domain Ω, σ_1 is a part of the boundary of the subdomain Ω_0, coincident with the straight line $x = 1$, σ_0 is the rest part of the boundary, $u = u(z)$ is a bounded solution of the boundary-value problem

$$\Delta u = 0 \quad \text{on} \quad \Omega_0, \qquad u = j \quad \text{on} \quad \sigma_j, \qquad j = 0, 1, \tag{28.2}$$

$v = v(z)$ is a harmonic function conjugate to u, with $v(a) = 0$.

It is obvious that the function

$$\zeta = \zeta(z) = u + iv \tag{28.3}$$

conformally maps the subdomain Ω_0 onto the strip

$$0 < \operatorname{Re} \zeta < 1.$$

Here σ_j is carried over to the straight line $\operatorname{Re} \zeta = j$, $j = 0, 1$, and the normalizing condition

$$\zeta(a) = 0 \qquad\qquad (28.4)$$

is satisfied.

The conformal mapping of subdomain (28.1) onto a strip possessing the indicated properties is unique.

Let us analytically continue function (28.3) from Ω_0 to the domain Ω by using the symmetry principle once again retaining the same notations for the continued function and its real and imaginary parts. The continued function (28.3) obviously conformally maps the domain Ω (the exterior of the lattice of ellipses) onto the domain ω which is the exterior of the lattice of plates (line segments)

$$\gamma_p = \{\zeta : \operatorname{Re} \zeta = 2p, \ |\operatorname{Im} \zeta| \le l\}, \qquad p = 0, \pm 1, \pm 2, \ldots,$$

where

$$l = v(ib) \qquad\qquad (28.5)$$

is a quantity to be defined, dependent on a and on b.

28.2.   Constructing Blocks

Suppose that $c^2 = b^2 - a^2$, A is a number, $1 < A \le 2 - a$, $B = (A^2 + c^2)^{1/2}$, $D = B(1 - A^{-2})^{1/2}$ is the ordinate of one of the points of intersection of the straight line $x = 1$ and an ellipse confocal with the ellipse Γ_0 and having semiaxes A and B. Below we only consider the case when $D > b$.

We specify a number D_1, $b \le D_1 < D$, and set

$$T_1 = \{z : 0 < x < 1, \ y > D_1\}.$$

Suppose also that

$$T_2 = \left\{ z : \frac{x^2}{a^2} + \frac{y^2}{b^2} > 1, \ \frac{x^2}{A^2} + \frac{y^2}{B^2} < 1 \right\}.$$

We call T_1 an extended block-half-strip and T_2 an extended block of the type of ring. We also specify a number D_2, $D_1 < D_2 < D$, and introduce basic blocks

$$T_1^* = \{z : 0 < x < 1, y > D_2\},$$

$$T_2^* = \Omega_0 \cap \{z : 0 < y < D_2\}.$$

We have

$$T_\mu^* \subset T_\mu \subset \Omega, \qquad (\overline{T}_\mu^* \cap \Omega_0') \subset T_\mu, \qquad \overline{T}_1^* \cup \overline{T}_2^* = \overline{\Omega}_0', \qquad (28.6)$$

where $\mu = 1, 2$,

$$\Omega_0' = \Omega_0 \cap \{z : y > 0\}. \qquad (28.7)$$

It is obvious that the function

$$w = f_1(z) \equiv \exp\{i\pi(z - iD_1)\} \qquad (28.8)$$

conformally maps T_1 onto the half-disk $|w| < 1$, Im $w > 0$, and the function

$$w = f_2(z) \equiv (z + \sqrt{z^2 + c^2})/(A + B) \qquad (28.9)$$

conformally maps T_2 onto the ring $h < |w| < 1$, where

$$h = (a + b)/(A + B). \qquad (28.10)$$

We denote by $f_\mu^{-1}(w)$ the inverse of the function $f_\mu(z)$ and set

$$\psi_\mu(\beta) = u(f_\mu^{-1}(e^{i\beta})), \qquad \mu = 1, 2, \qquad (28.11)$$

where u is the solution of the boundary-value problem (28.2) continued analytically from Ω_0 to Ω.

Suppose that

$$\genfrac{}{}{0pt}{}{R_1}{I_1}(z, \eta) = \sum_{p=0}^{1}(-1)^p \genfrac{}{}{0pt}{}{R}{I}\left(|f_1(z)|, \arg f_1(z), (-1)^p\eta\right), \qquad (28.12)$$

$$\genfrac{}{}{0pt}{}{R_2}{I_2}(z, \eta) = \sum_{p=0}^{1}\sum_{q=0}^{1}(-1)^q \genfrac{}{}{0pt}{}{R_0}{I_0}\left(|f_2(z)|, \arg f_2(z), \varphi_{pq}(\eta)\right), \qquad (28.13)$$

where R, I are conjugate kernels (3.14), (8.13),

$$\frac{R_0}{I_0}(r,\theta,\beta) = \frac{\mathrm{Re}}{\mathrm{Im}} \zeta_0(t,h),$$

h is quantity (28.10), $t = r\exp\{i(\theta - \beta)\}$, ζ_0 is Weierstrass' function (19.9), and

$$\varphi_{pq}(\eta) = (-1)^p(\pi(1 - (-1)^q)/2 - \eta).$$

On the basis of Theorem 3.1, the function u can be represented on $T_1 \cup \overline{T}_1^*$ as

$$u(z) = x + \int_0^\pi \left(\psi_1(\eta) - \frac{\eta}{\pi}\right) R_1(z,\eta)\, d\eta, \qquad (28.14)$$

where ψ_1 is function (28.11) for $\mu = 1$.

According to Villat's formula (see (19.4)), with due account of the evenness of the function u relative to the real axis and its oddness relative to the imaginary axis, this function can be represented on $T_2 \cup \overline{T}_2^*$ as

$$u(z) = \int_0^{\pi/2} \psi_2(\eta) R_2(z,\eta)\, d\eta, \qquad (28.15)$$

where ψ_2 is function (28.11) for $\mu = 2$.

28.3. The Algorithm

We choose a natural number n and specify points P_k, $k = 1, 2, \ldots, 2n$, with a complex coordinate

$$z_k = x_k + iy_k = \begin{cases} f_2^{-1}(ie^{i\pi(1-2k)/4n}), & 1 \leq k \leq n, \\ (2k-1)/2n - 1 + iD_2, & n < k \leq 2n. \end{cases} \qquad (28.16)$$

When $1 \leq k \leq n$ the points P_k lie on the boundary of the block T_2 and when $n < k \leq 2n$ they lie on the boundary of the block T_1.

We associate every point P_k with a point P_k' with a complex coordinate

$$z_k' = x_k' + iy_k' = \begin{cases} x_k + iy_k, & x_k \leq 1, \\ 2 - x_k + iy_k, & x_k > 1, \end{cases} \qquad (28.17)$$

and set

$$T_k = \begin{cases} 1, & z'_k = z_k, \\ -1, & z'_k \ne z_k. \end{cases} \qquad (28.18)$$

By virtue of (28.6), (28.16), (28.17) and with due regard for the fact that P_k, $P'_k \in \overline{\Omega}$ and $u(1 + iy) \equiv 1$, $|y| < \infty$, we have

$$u(z_k) = \begin{cases} u(z'_k), & z'_k = z_k, \\ 2 - u(z'_k), & z'_k \ne z_k \end{cases} \qquad (28.19)$$

for $1 \le k \le 2n$.

Let

$$n_0 = \min\{k : y_k \le D_2\} - 1,$$

and then

$$P'_k \in \begin{cases} \overline{T}^*_1, & 1 \le k \le n_0, \\ T^*_2, & n_0 < k \le 2n. \end{cases} \qquad (28.20)$$

Note that under the conformal mappings (28.8) and (28.9) of the extended blocks T_1 and T_2, the harmonic function $u = u(z)$ defined on them is carried over to the harmonic functions $u_1(w) = u(f_1^{-1}(w))$ and $u_2(w) = u(f_2^{-1}(w))$, defined on the half-disk $|w| < 1$, $\mathrm{Im}\, w > 0$ and on the ring $h < |w| < 1$, respectively.

Preserving the equality sign, we replace the integrals in (28.14) and (28.15) by the quadrature formulas of rectangles with n nodes which are, respectively, the images of the points P_k, $k = n + 1, n + 2, \ldots, 2n$, under mapping (28.8) and the images of these points for $k = 1, 2, \ldots, n$ under mapping (28.9). Then, specifying successively $z = z'_k$, $k = 1, 2, \ldots, 2n$, and, taking into account (28.6), (28.11), (28.16)–(28.20), and the remark made above, we arrive at the following system of linear algebraic equations for the approximate values u^k of the function u at the points P_k:

$$u^k = 1 - T_k$$

$$+ T_k \left(x'_k + \frac{\pi}{n} \sum_{m=1}^{n} \left(u^{n+m} - \frac{2m-1}{2n} \right) R_1 \left(z'_k, \pi \frac{2m-1}{2n} \right) \right), \qquad (28.21)$$

$$1 \le k \le n_0,$$

$$u^k = 1 - T_k + T_k \frac{\pi}{2n} \sum_{m=1}^{n} u^m R_2 \left(z'_k, \frac{\pi}{2} \left(1 - \frac{2m-1}{2n} \right) \right), \qquad (28.22)$$

$$n_0 < k \le 2n.$$

Let u_ν^k, $1 \le k \le 2n$, be an approximate solution of system (28.21), (28.22), obtained after the νth Seidel iteration at the initial approximation $u_0^k = 1.25$ for $1 \le k \le n$ and $u_0^k = (2k-1)/2n - 1$ for $n < k \le 2n$. Approximating the integrals in (28.14), (28.15) by the quadrature formulas of rectangles and replacing at the nodes, the unknown values of the functions ψ_1, ψ_2 by their approximate values u_ν^k, we get an approximate solution of the boundary-value problem (28.2) on the closed basic blocks $\overline{T}_1^*$, $\overline{T}_2^*$ in the form

$$u_{1n}^\nu(z) = x + \frac{\pi}{n} \sum_{k=1}^{n} \left(u_\nu^{n+k} - \frac{2k-1}{2n} \right) R_1\left(z, \pi \frac{2k-1}{2n} \right),$$

$$u_{2n}^\nu(z) = \frac{\pi}{2n} \sum_{k=1}^{n} u_\nu^k R_2\left(z, \frac{\pi}{2}\left(1 - \frac{2k-1}{2n} \right) \right),$$

respectively. The harmonic functions v_{1n}^ν, v_{2n}^ν conjugate to u_{1n}^ν, u_{2n}^ν have an explicit expression

$$v_{1n}^\nu(z) = C + y + \frac{\pi}{n} \sum_{k=1}^{n} \left(u_\nu^{n+k} - \frac{2k-1}{2n} \right) I_1\left(z, \pi \frac{2k-1}{2n} \right), \qquad (28.23)$$

$$v_{2n}^\nu(z) = \frac{\pi}{2n} \sum_{k=1}^{n} u_\nu^k I_2\left(z, \frac{\pi}{2}\left(1 - \frac{2k-1}{2n} \right) \right), \qquad (28.24)$$

where C is a constant.

In the process of iteration, when solving system (28.21), (28.22), we calculate the control quantity

$$\varepsilon_{n\nu}^0 = \max_{1 \le p \le 15} \left| u_{2n}^\nu \left(1 + i D_2 \frac{2p-1}{30} \right) - 1 \right|, \qquad (28.25)$$

characterizing the deviation on $\overline{T}_2^* \cap \sigma_1$ of the approximate solution u_{2n}^ν from the defined boundary value of u on σ_1 (see (28.2)). The iterations are performed up to the minimum value $\nu \ge 2$, for which condition (15.13) is fulfilled.

The approximate value l_n of quantity (28.5) can be found from the formula

$$l_n = v_{2n}^\nu(ib), \qquad (28.26)$$

and the constant C in expression (28.23) (when ν is fixed) is chosen on the basis of the requirement

$$v_{1n}^{\nu}(iD_2) = v_{2n}^{\nu}(iD_2)$$

with due regard for the fact that function (28.24) satisfies the condition

$$v_{2n}^{\nu}(x) \equiv 0, \qquad a \leq x \leq 1.$$

The approximate conformal mapping of subdomain (28.7) onto the half-strip

$$0 < \mathrm{Re}\,\zeta < 1, \qquad \mathrm{Im}\,\zeta > 0$$

is defined on the half-open basic blocks $\overline{T}_\mu^* \cap \Omega_0'$, $\mu = 1, 2$, by means of the functions

$$\zeta = \zeta_{\mu n}(z) = u_{\mu n}^{\nu}(z) + i v_{\mu n}^{\nu}(z). \tag{28.27}$$

These functions are pratically defined on the closed blocks $\overline{T}_\mu^*$, $\mu = 1, 2$, and assume close values at their intersection which is a line segment.

With the aid of the symmetry principle, the constructed approximate conformal mapping of subdomain (28.7) leads to an approximate conformal mapping of subdomain (28.1) onto the strip $0 < \mathrm{Re}\,\zeta < 1$, condition (28.4) being fulfilled. A further use of the principle of symmetry yields an approximate conformal mapping of the whole infinitely-connected domain Ω onto the exterior of the lattice of plates (line segments).

28.4. Practical Results

Table 28.1 gives the results of computations, carried out according to the presented algorithm, on the BESM-6 computer in the single-precision mode when the exterior of the lattice, formed by the ellipses with semiaxes $a = 0.2$ and $b = 1.2$, is mapped onto the exterior of the lattice of the line segments. We have chosen parameters $A = 1.8$, $D_1 = 1.38$, $D_2 = 1.5$, with $D \approx 1.8$. When calculating the series in (19.9), we retained the terms for which $h^{2k-2} > 2^{-45}$. Table 28.1 presents the values of the following quantities: n is the value of the

Table 28.1

n	ν	l_n	$\varepsilon^0_{n\nu}$
5	2	1.53	$7.5 \cdot 10^{-2}$
10	3	1.533 8	$8.2 \cdot 10^{-3}$
15	4	1.533 726	$4.8 \cdot 10^{-4}$
20	6	1.533 729 94	$1.9 \cdot 10^{-5}$
25	7	1.533 729 948	$2.4 \cdot 10^{-6}$

discreteness parameter, ν is the number of iterations carried out to solve system (28.21), (28.22), l_n is the approximate value of the required quantity (28.5) obtained from formula (28.26), $\varepsilon^0_{n\nu}$ is the indicator of accuracy (28.25).

Small values of ν suggest a rapid convergence of iterations. The observed growth of the coincident digits in the value of l_n with an increase in n and the behavior of $\varepsilon^0_{n\nu}$ substantiate the exponential convergence of the block method. The value of l_n with 10 digits found for $n = 25$ is reliable since it is repeated for a number of subsequent n. The computing time for one variant is of the order of 10 seconds.

The efficiency of the block method has been verified for the extreme cases, namely, for a lattice formed by circles of a sufficiently small radius or by circles of a radius close to unity, when only narrow gaps are between adjacent circles.

Assume now that $a = b$. The asymptotic formulas

$$l \sim 2a, \qquad a \to +0, \qquad \qquad (28.28)$$

$$l \sim \varkappa(a) = \frac{\pi}{\sqrt{2(1-a)}}, \qquad a \to 1 - 0, \qquad (28.29)$$

hold true, where l is quantity (28.5), a is the radius of the circles that form the lattice. We shall not derive the asymptotic formulas here.

Table 28.2 shows the way in which we can reach the asymptotic behavior (28.28) when carrying out real computations by the block method. The parameters that are not indicated in the table assume the following values: $n = 30$, $A = 1.5$, $D_1 = 0.5$, $D_2 = 0.8$.

For the values of a close to unity, the block T_2 of the type of ring is replaced by a block-bridge (see Sec.21). The results presented in Table 28.3 correspond to the asymptotic behavior (28.29). The value of l_n is given here with only two digits with the only purpose to demonstrate its rapid growth as $a \to 1 - 0$. Recall that the step of the lattices is fixed and equals 2. Thus the quantity $1 - a$ is equal to half the width of the gap between two adjacent circles that form the lattice being mapped. These computations were carried out for $n = 10$, $D_1 = 1.25$, $D_2 = 1.95$.

The block method can also be used to perform an approximate conformal mapping of more complicated lattices formed, in particular, by circular polygons. The blocks used for circular polygons are described in [67].

Table 28.2

a	$l_n/2a$	ν	$\varepsilon_{n\nu}^0$
$0.5 \cdot 10^{-1}$	1.25	7	$2.9 \cdot 10^{-9}$
$0.5 \cdot 10^{-2}$	1.002 1	8	$2.9 \cdot 10^{-9}$
$0.5 \cdot 10^{-3}$	1.000 021	8	$2.9 \cdot 10^{-9}$
$0.5 \cdot 10^{-6}$	1.000 000 000	8	$2.9 \cdot 10^{-9}$

Table 28.3

$1 - a$	l_n	$l_n/\varkappa(a)$	ν	$\varepsilon_{n\nu}^0/l_n$
10^{-1}	$5.4 \cdot 10^0$	0.77	4	$4.8 \cdot 10^{-9}$
10^{-3}	$6.9 \cdot 10^1$	0.98	5	$4.9 \cdot 10^{-11}$
10^{-7}	$7.0 \cdot 10^3$	0.994	5	$2.6 \cdot 10^{-12}$
10^{-10}	$2.2 \cdot 10^5$	0.999 98	4	$2.5 \cdot 10^{-12}$
10^{-14}	$2.2 \cdot 10^7$	0.999 999 91	3	$1.2 \cdot 10^{-12}$

References

1. N. I. Akhiezer, *Elements of the theory of elliptic functions*, Nauka Publ., Moscow, 1970; English transl. in Amer. Math. Soc., Providence, R.I., 1990.
2. M. A. Aleksidze, *Solving boundary-value problems by the method of expansion in terms of nonorthogonal functions*, Nauka Publ., Moscow, 1978 (Russian).
3. J.-P. Aubin, *Approximation of elliptic boundary-value problems*, Wiley-Interscience, New York–London–Sydney–Toronto, 1972.
4. K. I. Babenko, *Fundamentals of numerical analysis*, Nauka Publ., Moscow, 1986 (Russian).
5. I. Babuška, M. Práger and E. Vitásek, *Numerical processes in differential equations*, Wiley-Interscience, London–New York–Sydney, 1966.
6. N. S. Bakhvalov, N. P. Zhidkov and G.M. Kobel'kov, *Numerical methods*, Nauka Publ., Moscow, 1987 (Russian).
7. I. S. Berezin and N. P. Zhidkov, *Computing methods*. Vol. II, Fizmatgiz, Moscow, 1960; English transl., Addison-Wesley, Reading, Mass.; Pergamon Press, New York, 1965.
8. F. Bowman, *Notes on two-dimensional electric field problems*, Pros. London Math. Soc. (2) **39** (1935), 205–215.
9. L. Collatz, *Numerische Behandlung von Differentialgleichungen* (2. Aufl.), Springer-Verlag, Berlin–Göttingen–Heidelberg, 1955.
10. E. G. D'yakonov, *Difference methods for solving boundary-value problems*, Issue 1 (Stationary problems), Moscow Univ. Press, 1971 (Russian).
11. E. G. D'yakonov, *Minimization of computational work*, Nauka Publ., Moscow, 1989 (Russian).
12. P. F. Fil'chakov, *Approximate methods of conformal mappings*, Naukova Dumka Publ., Kiev, 1964 (Russian).
13. V. P. Fil'chakova, *Conformal mappings of a special type of domains*, Naukova Dumka Publ., Kiev, 1972 (Russian).
14. G. Forsythe and W. Wasow, *Finite-difference methods for partial differential equations*, Wiley, New York–London, 1960.
15. D. Gaiėr, *Konstruktive Methoden der konformen Abbildung*, Springer–Verlag, Berlin–Göttingen–Heidelberg, 1964.
16. S. K. Godunov and V. S. Ryaben'kiĭ, *Difference schemes*, Nauka Publ., Moscow, 1973; English transl. in North Holland, Amsterdam–New York–Oxford–Tokyo, 1987.
17. G. M. Goluzin, *Geometric theory of functions of a complex variable*, Nauka Publ., Moscow, 1966; English transl. in Amer. Math. Soc., Providence, R.I., 1969.
18. D. M. Hough and N. Papamichael, *The use of splines and singular functions in an integral equation method for conformal mapping*, Numer. Math. **37** (1981), 133–147.
19. L. V. Kantorovich and V. I. Krylov, *Approximate methods of higher analysis*, 3rd ed., GITTL, Moscow, 1950; English transl. in Interscience, New York, and Noordhoff, Groningen, 1958.

20. M. V. Keldyš, *On the solvability and stability of the Dirichlet problem*, Uspehi Mat. Nauk **8** (1941), 171–231; English transl. in Amer. Math. Soc. Transl. (2) **51** (1966).

21. P. Koebe, *Abhandlungen zur Theorie der konformen Abbildung*. IV, Acta Math. **41** (1916/18), 305–344.

22. W. Koppenfels and F. Stallmann, *Praxis der konformen Abbildung*, Springer-Verlag, Berlin–Göttingen–Heidelberg, 1959.

23. M. A. Lavrent'ev and B. V. Šabat, *Methods of the theory of functions of a complex variable*, 4th ed., Nauka Publ. Moscow, 1973; German transl. of 3rd ed. VEB Deutscher Verlag, Berlin, 1967.

24. V. N. Malozemov, *Estimate of the accuracy of a quadrature formula for periodic functions*, Vestnik Leningrad. Univ. (Ser. Mat. Mech. Astr.), No. 1 (1967), 52–59 (Russian).

25. G. I. Marchuk and V. I. Agoškov, *Introduction into projection-net methods*, Nauka Publ., Moscow, 1981; French. transl., Mir Publ., Moscow, 1985.

26. G. I. Marchuk and Yu. A. Kuznetsov, *Iteration methods and quadratic functionals*, Nauka Publ., Novosibirsk, 1975 (Russian).

27. G. I. Marchuk and V. I. Lebedev, *Numerical methods in the theory of neutron transport*, Atomizdat, Moscow, 1981; English transl., Harwood Acad. Publ., Chur, Switzerland, 1986.

28. G. I. Marchuk and V.V. Šhaidurov, *Increasing the accuracy of solutions of difference schemes*, Nauka Publ., Moscow, 1979; English transl., Springer-Verlag, New York–Berlin–Heidelberg–Tokyo, 1983.

29. V. P. Mikhailov, *Partial differential equations*, Nauka Publ., Moscow, 1976; English transl., Mir Publ., Moscow, 1978.

30. S. G. Mikhlin, *Numerical realization of variational methods*, Nauka Publ., Moscow, 1966; English transl., Wolters-Noordhoff, Groningen, 1971.

31. S. G. Mikhlin, *Errors in computation processes*, Tbilisi: Izd. Tbilisi Univ., 1983 (Russian).

32. A. Mitchell, and R. Wait, *The finite element method in partial differential equations*, Wiley, New York–Brisbane–Toronto, 1977.

33. L. A. Oganesyan and L. A. Rukhovets, *Variational-difference methods of solving elliptic equations*, Erevan: Izd. Akad. Nauk Armen. SSR, 1979 (Russian).

34. N. Papamichael N. and C. A. Kokkinos, *The use of singular functions for the approximate conformal mapping of doubly-connected domains*, SIAM J. Sci. Stat. Comput. **5** (1984), 684–700.

35. I. G. Petrovskiĭ, *Partial differential equations*, 3rd ed., Fizmatgiz, Moscow, 1961; English transl., Saunders, Philadelphia, Pa., 1967.

36. V. S. Ryabenkiĭ, *Difference potentials method for some problems of mechanics of continuum*, Nauka Publ., Moscow, 1987 (Russian).

37. V. S. Ryabenkiĭ and A. F. Fillippov, *On the stability of difference equations*, GITTL, Moscow, 1956; German transl., Mathematik für Naturwiss. und Technik, Band 3, VEB Deutscher Verlag, Berlin, 1960.

38. A.A. Samarskiĭ, *Theory of difference schemes*, Nauka Publ., Moscow, 1977 (Russian).

39. A. A. Samarskiĭ and V. B Andreev, *Difference methods for elliptic equations*, Nauka Publ., Moscow, 1976; French transl., Mir Publ., Moscow, 1978.

40. A. A. Samarskiĭ and A. V. Gulin, *Stability of difference schemes*, Nauka Publ., Moscow, 1973 (Russian).

41. A. A. Samarskiĭ, R. D. Lazarov and V. L. Makarov, *Difference schemes for differential equations with generalized solutions*, Vyssh. Shkola Publ., Moscow, 1987 (Russian).

42. A. A. Samarskiĭ and E. S. Nikolaev, *Methods of solving net equations*, Nauka Publ., Moscow, 1978; English transl., Birkhäuser Verlag, Basel–Boston–Berlin, 1989.

43. V. K. Saulyev, *Integration of equations of parabolic type by the method of nets*, Fizmatgiz, Moscow, 1960; English transl., Pergamon Press, New York, 1964.

44. M. Schiffer, *Some recent developments in the theory of conformal mapping*, Appendix to R. Courant, *Dirichlet's principle, conformal mapping, and minimal surfaces*, Interscience (1950), 249–323.

45. S. L. Sobolev, *Applications of functional analysis in mathematical physics*, Novosibirsk: Izd. Sib. Otd. Akad. Nauk SSSR, 1962; English transl. in Amer. Math. Soc., Providence, R.I., 1963.

46. G. Strang and G. Fix, *An analysis of the finite element method*, Englewood Cliffs (USA): Prentice-Hall, 1973.

47. V. I. Vlasov, *Boundary-value problems in domains with curvilinear boundaries*, Izd. Comp. Centre Akad. Nauk SSSR, Moscow, 1987 (Russian).

48. E. A. Volkov, *On the solution by the finite difference method of the inner Dirichlet problem for the Laplace equation*, Vychisl. Mat. 1 (1957), 34–61, Izd. Akad. Nauk SSSR; English transl. in Amer. Math. Soc. Transl. (2) 24 (1963).

49. E. A. Volkov, *Differentiability properties of solutions of boundary-value problems for the Laplace and Poisson equations on a rectangle*, Trudy Mat. Inst. Steklov, 77 (1965), 89–112; English transl. in Proc. Steklov Inst. Math. 77 (1965).

50. E. A. Volkov, *Differentiability properties of solutions of boundary-value problems for the Laplace equation on a polygon*, Trudy Mat. Inst. Steklov, 77 (1965), 113–142; English Transl. in Proc. Steklov Inst. Math, 77 (1965).

51. E. A. Volkov, *A finite difference method for finite and infinite regions with piecewise smooth boundary*, Dokl. Akad. Nauk SSSR, 168, 5 (1966), 978–981; English transl. in Soviet Math. Dokl. 7 (1966).

52. E. A. Volkov, *The method of composite nets for finite and infinite regions with piecewise smooth boundary*, Trudy Mat. Inst. Steklov, 96 (1968), 117–148; English transl. in Proc. Steklov Inst. Math., 96 (1968).

53. E. A. Volkov, *The method of regular composite nets for solving the mixed boundary-value problem for the Laplace equation*, Dokl. Akad. Nauk SSSR, 196, 2 (1971), 266–269; English transl. in Soviet Math. Dokl. 12 (1971).

54. E. A. Volkov, *On the regular composite difference method for Laplace's equation on polygons*, Trudy Mat. Inst. Steklov, 140 (1976), 68–102; English transl. in Proc. Steklov Inst. Math. 140, 1 (1979).

55. E. A. Volkov, *A difference-analytic method for calculating the potential field on polygons*, Dokl. Akad. Nauk SSSR, **237**, 6 (1977), 1265–1268; English transl. in Soviet Math. Dokl. **18** (1977).

56. E. A. Volkov, *A rapidly converging method of quadratures for solving Laplace's equation on polygons*, Dokl. Akad. Nauk SSSR, **238**, 5 (1978), 1036–1039; English transl. in Soviet Math. Dokl. **19** (1978).

57. E. A. Volkov, *An exponentially converging method for solving Laplace's equation on polygons*, Mat. Sb. **109**, 3 (1979), 323–354; English. transl. in Math. USSR Sb. **37** (1980).

58. E. A. Volkov, *An exponentially converging method of conformal mapping of polygonal regions*, Dokl. Akad. Nauk SSSR, **249**, 6 (1979), 1292–1295; English transl. in Soviet Math. Dokl. **20** (1979).

59. E. A. Volkov, *An exponentially converging method for the Neumann problem on multiply connected polygons*, Trudy Mat. Inst. Steklov, **172** (1985), 86–106; English transl. in Proc. Steklov Inst. Math. **172**, 3 (1987).

60. E. A. Volkov, *An approximate method for conformal mapping of multiply connected polygons onto canonical domains*, Trudy Mat. Inst. Steklov, **173** (1986), 55–68; English transl. in Proc. Steklov Inst. Math. **173**, 4 (1987).

61. E. A. Volkov, *Approximate conformal mapping of certain polygons onto a strip by the block method*, Zh. Vychisl. Mat. i Mat. Fiz. **27**, 8 (1987), 1166–1175; English transl. in USSR Comput. Math. and Math. Phys. **27**, 4 (1987), 136–142.

62. E. A. Volkov, *Development of the method of quadratures for the Laplace equation and for conformal mappings*, Trudy Mat. Inst. Steklov, **180** (1987), 83–85; English transl. in Proc. Steklov Inst. Math. **180**, 3 (1989).

63. E. A. Volkov, *Approximate conformal mapping of a square frame onto an annulus by the block method*, Investigations in the Theory of Approximation of Functions, Otdel. Fiz.–Mat. Bashkir. Filial Akad. Nauk SSSR, Ufa (1987), 85–96 (Russian).

64. E. A. Volkov, *High-precision practical results in conformal mappings of simply connected and doubly connected domains by the block method*, Trudy Mat. Inst. Steklov, **181** (1988), 40–69; English transl. in Proc. Steklov Inst. Math. **181**, 4 (1989).

65. E. A. Volkov, *Approximate conformal mapping by the block method of a circle with a polygonal hole into a ring*, Zh. Vychisl. Mat. i Mat. Fiz., **28**, 6 (1988), 835–841; English transl. in USSR Comput. Math. and Math. Phys. **28**, 3 (1988), 143–147.

66. E. A. Volkov, *Approximate conformal mapping of the exterior of a parabola with a hole onto an annulus*, Ukr. Mat. Zh. **41**, 4 (1989), 475–479; English transl. in Ukrain. Math. J. **41** (1989).

67. E. A. Volkov, *Development of the block method for solving the Laplace equation for finite and infinite circular polygons*, Trudy Mat. Inst. Steklov, **187** (1989), 39–68; English transl. in Proc. Steklov Inst. Math. **187**, 3 (1990).

68. E. A. Volkov, *Approximate conformal mapping of the exterior of a lattice of ellipses on the exterior of a lattice of plates by the block method*, Trudy Mat. Inst. Steklov, **192** (1990), 35–41; English transl. in Proc. Steklov Inst. Math. **192**, 3 (1992).

69. E. A. Volkov, *Approximate conformal mapping of domains with periodic structure by the block method*, Trudy Mat. Inst. Steklov, **200** (1991), 100–113 (Russian).

70. E. A. Volkov, *Approximate solution of the Laplace equation on polygons under analytic mixed boundary conditions by the block method*, Trudy Mat. Inst. Steklov, **201** (1992), 165–185 (Russian).

71. E. A. Volkov, *Approximate solution of the Laplace equation on polygons under nonanalytic boundary conditions by the block method*, Trudy Mat. Inst. Steklov, **194** (1992), 63–88 (Russian).

72. E. A. Volkov, *Rapid block method of constructing Green's function of the Laplace operator on polygons*, Dif. Ur. **28**, 7 (1992), 1189–1197 (Russian).

Index

Milton Keynes UK
Ingram Content Group UK Ltd.
UKHW031148141024
449569UK00024B/971